RIVERSIDE COMMUNITY COLLEGE
1916

Genetics & Society

Genetics & Society

BARRY HOLLAND
Université de Paris xi

CHARALAMBOS KYRIACOU
University of Leicester

Addison-Wesley Publishing Company
Wokingham, England • Reading, Massachusetts • Menlo Park, California
New York • Don Mills, Ontario • Amsterdam • Bonn • Sydney • Singapore
Tokyo • Madrid • San Juan • Milan • Paris • Mexico City • Seoul • Taipei

arlow, Buckinghamshire;
incorporating abstract photograph of head © Larry Keenan Assoc/The Image Bank and computer graphics image of DNA © Oxford Molecular Biophysics Laboratory/Science Photo Library. DNA (deoxyribonucleic acid) consists of two linked strands of nucleotides (blue and red), coiled into a helix. The strands are joined by the hydrogen bonding of pairs of organic bases (purple and orange) on adjacent nucleotides. The organic bases are thymine, adenine, cytosine and guanine. The green circles show where water is likely to bind on the outside of the DNA molecule. Water serves a number of functions and is involved in the interaction of DNA with proteins. The image shows the β-form of DNA (β-DNA).
Cover printed by The Riverside Printing Co. (Reading) Ltd.
Typeset by Colset Pte Ltd, Singapore in Plantin
Printed in Great Britain at the University Press, Cambridge

First printed 1993

ISBN 0-201-56515-3

British Library Cataloguing in Publication Data
A cataloque record for this book is available from the British Library

Library of Congress Cataloging in Publication Data
Genetics and society / [edited by] Barry Holland, Charalambos Kyriacou.
p. cm.
Includes index.
ISBN 0-201-56515-3 : £0.26
1. Human genetics–Social aspects. 2. Genetic engineering–Social aspects. 3. Medical genetics–Social aspects.
QH431.G4324 1993
303.48′3–dc20 93-13618
CIP

Acknowledgement

The publishers would like to thank the following for permission to reproduce material in this book. Article by Michael Rennie 'I'm the only one who knows how much it hurts' in *The Independent*; Stent G.S. (1971), *Molecular Genetics: An Introductory Narrative* reprinted with permission of W.H. Freeman and Company; Weatherall D.J. (1991) *The New Genetics and Clinical Practice* Oxford University Press by permission of Oxford University Press; Chamberlain J.S. *et al.* (1990), in: PCR Protocols: A Guide to Methods and Applications (Innis M. *et al.* eds) Academic Press; McLaren A. (1987), 'Can we diagnose genetic disease in pre-embryos?' *New Scientist*; McLaren A. (1990). Research on the human concepts and its regulation in Britain today, *Journal Royal Society of Medicine*; Watson J., Tooze J. and Kurtz D.T. (1983), *Recombinant DNA. A short course* reprinted with permission of W.H. Freeman and Company; Rossiter B.J.F. and Caskey C.T. (1991), 'Molecular studies of human genetic disease, *FASEB journal*; McLaren A. (1984), '*Where to draw the line*' Royal Institution of Great Britain.

Preface

A visitor from Cygnus X, seeing our wealth of accumulated scientific knowledge and literary masterpieces, would be dismayed to view the inequalities of life on our planet; the excesses of food production in one quarter mirrored by total deprivation in another; high-tech molecular dissection of the AIDS virus (our modern-day Black Death) for the benefit of Westerners and the lack of the simplest preventive measures in the underdeveloped countries where the disease spreads unchecked. Fuelled by arms manufacturers (acting in our name), wars rage, based on the most grotesque excuses of tribalism, nationalism and narrow religious convictions. Perhaps Koestler fingered our predicament correctly: we are fundamentally flawed. Our genetic blueprint, an amalgam of our reptilian origin and the explosive, runaway development of our higher order brain centres, is destined for self-destruction. Fortunately, whether designed by a god to test us, or originating as a random flux of cosmic matter, life's genetic programme in all its glittering detail is within our grasp, only seconds away on the Earth's vast time-scale. Within our lifetime, the opportunity will be there, if we can exploit it, to raise humanity to a far higher plane than hitherto. This has been the objective of all societies, but never has such an opportunity faced a previous generation.

Scientists will initially be in the front line of this exploration and perhaps we should consider that, faced with such a responsibility, they should be better and more broadly educated in the future, better able to grasp the social and ethical implications of scientific advances. Active or former scientists should surely participate much more fully in commerce, industry and in politics at all levels, including the highest levels of government, if society is truly to benefit from the pursuit and application of knowledge.

It is a plain fact that many scientists have been guilty of exaggeration, if not wild fantasy, in their predictions of the imminent benefits of biotechnology. This prevailing habit is often a response to the international trend of governments demanding justification for every dollar spent on fundamental research. Scientists are human beings, and to protect their way of life, the pursuit of knowledge, they may allow their imaginations to overheat in order to justify fundamental research. We all lose as a result, and governments must relax this pressure, which as we shall outline later would be far better employed in other ways. A return to a better balance of the funding of pure research, and perhaps a more imaginative way of promoting applied research in public institutions, must be sought. At the same time, scientists must try harder to temper their fantasies, and the rest of us must be more circumspect about their frothier claims for the relevance of their work.

The nub of our present mismanagement of science in virtually all countries is, however, the total failure of scientists and society's leaders to have any kind of meaningful dialogue. When assessing cost and benefit of science at company, national or international level, the absence of an informed scientific input can have catastrophic consequences. The gulf between scientists and society's leaders is now causing, in our view, a particular problem in relation to the accumulating knowledge of the human genome, introduced in this volume by James Watson. Paradoxically, in relation to prospects for genetic (some would say social) engineering, through sheer defensiveness scientists are in grave danger of actually playing down the implications of the rapid advances now being made towards the total elucidation of the structure of genomes (the genetic blueprints) of several bacterial species, yeast, fruit flies, plants and the human genome itself. These advances, and the technology for the manipulation of genes (which is now largely in place), are being minimized to the point where society will not be alert to the coming consequences. Gene therapy for diseases such as cystic fibrosis is being tested in humans as we go to press, pre-implantation diagnosis in countries such as the UK is well established, and landmark developments in agricultural practice – the release of genetically engineered food crops for human consumption – will occur in the USA in 1993.

Let us be perfectly clear and unequivocal: we are now on an unstoppable path to an understanding of the nature and properties of every one of the 100 000 or more human genes, combined with (even using present day technology), the ability to isolate any human, plant or animal gene, to produce thousands of variants in the test tube in a few hours, and to transfer in a trice any gene or its variants from any one organism to another, from humans to bacteria. Without doubt, we have a revolution on our hands which we must guide skilfully and sensitively if we are to exploit this successfully. Whether the final understanding of how all our genes work takes 30 or 50 years, we cannot now escape the inevitable, ultimate result. We shall, for example, understand the developing, genetically determined programme of heart disease, we shall understand the basis of intelligence and behaviour, and we shall, if we so wish, interfere with these God-given programmes, for our perceived benefit, at the single egg cell, embryonic or adult stages of human development. The Jeremy Rifkins of this world are perfectly correct in pointing out the worst scenarios, but unlike these critics we recognize that the tide cannot be turned and we must work to understand and deal with these revolutionary changes to the human and other biological systems on this planet. This can only be achieved by involving the whole of society in discussing how human life will henceforth evolve, for the first time not only as the prerogative of a god or nature, but increasingly under *our* human control.

The revolutionary developments now envisaged, including the manipulation of human genes to eradicate hereditary disease, rational design of new drugs and the genetic engineering of all-purpose plants, will as usual largely pass the developing world by, if indeed they do not hasten its further decline. To avoid this, ways must be found to motivate companies to invest in the development and distribution of ever more sophisticated drugs and modified crops, specifically for use in Third World countries. This should involve subsidies from Western governments to industry, for example in the form of purchasing credits in lieu of aid to underdeveloped countries. In addition, the promotion in developed countries of a cadre of publicly funded applied scientists, as opposed to purely academic scientists, within the existing academic community, could facilitate a more enlightened development and application of science and technology than we have seen so far. This applies to Third World assistance, but equally to stimulating technological interchange between the private and public scientific communities in the advanced countries.

In our view, the application of modern advances in genetics, whether medical or agricultural, cannot be allowed to develop *ad hoc*. Neither should it be unnecessarily constrained by latter-day Luddites. Clear leadership will be required from governments acting through ethical and other advisory committees with representatives from all sectors of society, including scientists. Whether the resulting guidelines should be voluntary or mandatory will be determined by circumstances and the democratic process, and one view on how we might deal with the new knowledge is considered by Robert Pritchard in this volume. We live in a society dominated by science, but few governments have a scientific policy and politicians with any scientific background are rare birds indeed. This must surely change in relation to all applications of scientifically based technologies, not least the utilization of genetic engineering. Governments must be prepared to formulate high technical standards of manufacture and professional practice, with which national and international companies and other organizations, competing to meet the same standards, must comply. Such standards can include those for drug design, modified crops, novel pesticides and herbicides, gene therapy and procedures for the elimination of defective genes. Fortunately, it seems that we are emerging from a period when some Western governments have abjectly forsaken their real responsibilities in the name of non-intervention in the so-called market, which has expanded to all aspects of our society. We rightly expect our politicians, democratically elected, to be more imaginative and to intervene to ensure that high standards of technological application are adhered to, and to actively promote education and health care for all citizens. If developments in genetic screening procedures are not to be abused, governments will have to work with and, if necessary, underpin the insurance industry in order to safeguard individuals with 'high risk' genetic traits so that they are not disadvantaged. The public interest will require to be much more vigorously represented, for example, by a 'public scrutineer' at the very outset of new patent applications. It is not in the public interest to foreclose access to data accruing from the human genome project on the basis of patent claims.

In the face of the challenging period ahead, with Pandora's box fully open and our genetic blueprint laid out before us, we remain optimistic that the new knowledge can be beneficially managed, provided we meet head on its implications through continuing debate at all levels of society. The prospects are immense: human society can be transformed and we should not fear the future. However, the foundations must be laid now for the full and sensible exploitation of our genetic inheritance by future generations.

In this book we have brought together world leaders in genetic engineering in relation to agriculture, forensic medicine, the human genome, the treatment of human disease and the elimination of hereditary disease through pre-implantation diagnosis. We have asked them to describe in the broadest terms possible what they do and what they hope to achieve in the name of science. The result is a clear exposition of the basis of current technology, the issues surrounding its implementation, spiced with real-life accounts of scientific discovery in true detective style. In addition, we asked a philosopher, Mary Warnock, for her views on the implications of the revolution in genetics. Baroness Warnock identifies some of the major moral, legal and ethical questions facing us, and includes her firm views on the manipulation of human eggs or sperm: the engineering of the *next* and *future* generations.

The production of this book was stimulated by the conviction that science and scientists must become more open, better able to explain aspirations and applications, indeed to integrate and be integrated fully into society. This volume represents a modest step towards a much larger objective, which in our view must see scientists being brought in from the cold and cease being the unnamed back-room eccentrics. We are proud to confirm that scientists like the rest of us come in all shapes and sizes, are both male and female, and have a wide spectrum of political and religious beliefs. We are delighted that a representative sample, Richard Flavell, Alec Jeffreys, Robert Pritchard, James Watson, Anne McLaren and Thomas Caskey were, together with Mary Warnock, able to accept our invitation to contribute to the original public meeting held in the De Montfort Hall, Leicester, and to commit their thoughts and hopes on paper for this book.

Barry Holland
Charalambos Kyriacou
June 1993

Introduction

The Department of Genetics in Leicester first saw the light of day in the academic year 1964–65. The Head of Department, Bob Pritchard, was one of the youngest professors ever appointed to a British chair of genetics and through the 1960s and 1970s his leadership made Leicester an internationally acclaimed centre of excellence for genetics and molecular biology, primarily in studies of lower organisms such as viruses, bacteria, fungi and moulds. In 1977 Alec Jeffreys joined the department, and by 1985 had become a household name as the discoverer and developer of genetic fingerprinting. This molecular technique has revolutionized forensic science, as well as other, less applied, but equally worthy fields of endeavour within molecular biology. Alec was one of the first in the department to work with mammals and higher organisms, but by 1990 nearly half its members were working with either flies, mice or humans, and the department maintained an extremely high profile in genetics and molecular biology. Most of the credit for the department's success in the 1980s, however, must go to Barry Holland who was appointed to the chair in 1985, succeeding Clive Roberts, who had taken over from Bob for a couple of years.

The academic year 1989–1990 was the 25th anniversary of the founding of the department and I, as a junior member of staff, thought that we might throw a party: cheese and biscuits, some beer and skittles, maybe a disco. I have simple tastes. Barry had another idea. Why not have an international conference in Leicester and invite top geneticists and molecular biologists from all over the world to speak to the general public about the important ethical values raised by the modern molecular revolution? We could raise money through sponsorship and might attract as many as 2000 people to the De Montfort Hall in the centre of Leicester. Barry always had a tendency for the big idea, and my first response was to tell him to lie down until this foolish impulse had passed. However, it was not to be, and Barry, our departmental administrator Margaret Peake and I started thinking of how to raise the money to finance this risky enterprise. We soon discovered that we simply did not have time to focus on advertising, raising money, dealing with the press, designing a stage, getting together a glossy programme, organizing lunch for 2000 people, arranging first aid and lost property facilities, crèches and so on. The meeting was saved when Bridget Seddon and her team from Waltham Business Services walked into our lives and took over almost all the administration. Barry, Margaret and I raised money from the Science and Engineering Research Council, the Agricultural and Food Research Council, Wellcome Biotech, the Imperial Cancer Research Fund, Cellmark Diagnostics Ltd, ICI plc, Legal and General Assurance Ltd, Glaxo Group Research Ltd, Dalgety, Spillers plc and the Medical Research Council. I lost count of the number of begging letters that we sent out, but I am extremely grateful to the institutions that financed us.

Who would speak at this meeting? An obvious choice was Alec Jeffreys, and after a few well-chosen threats he agreed to speak. Bob Pritchard, our founding father, is very active in local politics and we thought he would be provocative. After all, if we asked him how could he refuse? Two down, who else? Molecular biology really began in 1953 when Jim Watson and Francis Crick published their famous paper on the structure of the double helix. Bob Pritchard knew Jim Watson and wrote to him. Watson agreed to come. One of the best-known medical molecular geneticists is Tom Caskey, who works in Houston, USA. We approached him and were amazed by how enthusiastically he responded. How about Anne McLaren? Her unit in London had developed the pre-implantation DNA analysis of early embryos based on the pioneering work of Marilyn Monk (an ex-Leicester department member). We asked Anne and she agreed to come.

The impact of molecular genetics on agriculture requires serious consideration and we asked Richard Flavell, the best-known agro-molecular biologist in the UK. He, too, agreed to speak. Finally, we wanted an out-and-out ethics type, and the person who readily sprang to mind was Baroness Mary Warnock, chairperson of the Warnock Committee, which studied the use of embryos for medical research. She too said that she would be delighted to come. Thus we had brought together all the pieces of the puzzle and set about organizing them into some kind of picture of modern molecular genetics and the moral and ethical issues which surround it. Finally, we asked Sir George Porter (now Lord Porter), a Nobel laureate and at that time President of the Royal Society, to chair the meeting. As he was also the Chancellor of the University of Leicester he was happy to help us out, as was Ken Edwards, our Vice-Chancellor, and also a former geneticist.

So the day came (7 April 1990). We were understandably nervous that the general public might not be as interested in turning up as we had imagined. However, they appeared in their hundreds, paying a very modest entry fee thanks to our sponsors. The De Montfort Hall was almost full. We were pleasantly surprised and Sir George was amazed, especially as we were competing with Nottingham Forest *versus* Tottenham Hotspur just up the road. Needless to say, the speakers were terrific and the audience responded to them enthusiastically. The press was everywhere, the meeting was written up in all the national dailies, and our speakers did radio, TV and press interviews all day.

Bob Pritchard's contribution focused on the education of the public. After all, the decisions that have to be made on the uses of molecular genetics are much too important to be left solely to politicians. The technical power of molecular biology to improve the quality of life was tackled by Jim Watson, Alec Jeffreys, Tom Caskey and Anne McLaren. Its benefits and dangers to agriculture were discussed by Dick Flavell. Mary Warnock focused on ethical considerations. After the dust had settled and our celebratory party on Saturday night was over, Barry had another of his bright ideas in the early hours of the morning. Why not get these speakers to write a chapter based on their talks and publish the results as a collection of essays that would be understandable to the general public? I could think of a dozen objections: we would need to persuade a publisher to take this project, we would have to get someone to edit it, and, of course, we would also have to talk the speakers into writing a chapter, which is something they were all initially very reluctant to do. However, by flattery ('it is so rare for serious and brilliant scientists to write articles that are comprehensible to educated laymen'), appealing to sense of duty ('you will be performing a valuable public service') and well-timed threats not to pay their expenses for the meeting until the speakers agreed, they all, to a wo(man), said they would contribute. So my thanks (and Barry's) go to our seven authors for their magnificent contributions, both verbal and written. Finally, we must thank

Bridget Seddon from Waltham Business Services and her team, who took on this risky financial enterprise with skill, daring and wit. They may not have made a lot of money out of it, but we hope they gained as much intangible reward as we did. We could never have done it without them.

Over the 18 months following the meeting the chapters came in and were amended, boxes were added to explain molecular jargon and so on. The finished product is not a polished review of the entire field or of all the technical and ethical problems – it is essentially a record of what was said during that memorable and exhausting day. We hope that this will be of interest to those who are not scientists or philsophers, but who nevertheless wish to share in some of the excitement that is generated by the new genetics. For those with more specific interests, where appropriate, brief editorial notes have been added to incorporate the most up-to-date information.

Charalambos Kyriacou
June 1993

Contents

1

Public Participation in the Scientific Adventure

Professor Robert Pritchard

Prior to his appointment as Professor of Genetics at the University of Leicester, Robert Pritchard had been a member of scientific staff at the Medical Research Council's microbial genetics research unit at Hammersmith Hospital (1959–63), following which he spent a period as visiting research associate in the department of physics, Kansas State University.

The newly-established Department of Genetics at the University of Leicester was planned as part of a new concept, the teaching of an integrated degree course in biological sciences. Robert Pritchard played an influential role in the establishment of what has become one of the university's most outstanding achievements, the School of Biological Sciences. He was chairman of the school for two years, from October 1977 to August 1979, and was involved in the establishment of the Medical School in 1975, playing an important role in designing the pre-clinical course, of which genetics forms part. At the same time he led a large and successful research group, and spent several periods of study leave abroad, where he was much in demand as a research collaborator.

Since relinquishing the headship of the department in 1983, he has been able to devote more time to his interest in local politics.

A life-long supporter of the Liberal Party, he was elected to Leicester City Council as a Liberal-Alliance candidate and then to Leicestershire County Council as a Liberal Democrat.

Public misunderstanding of, and antipathy to, science and its exploitation is perverse in view of its unique potential to mitigate the environmental impact of human exploitation of world resources, and its potential to liberate the poor and dispossessed.

It is widely assumed that lack of education is the problem. I suggest that the scientific and political establishments must accept some responsibility.

The impressive intellectual structure erected by science over two centuries, and its technological spin-offs, have transformed people's lives and their thinking for the better. The application of knowledge to human affairs has brought wealth, health, liberty, understanding and justice disproportionately to the poor and the weak. The wealth it has generated has progressively extended to everyone the privileges, such as education and medicine and civil rights previously enjoyed by a wealthy minority at least in the richer parts of the world.

Recent advances in molecular biology are simply a continuation of this process. Richard Flavell's article in this symposium vividly illustrates how DNA technology can increase food production, lower its cost and reduce the damaging effects of the current crude use of pesticides and herbicides on the environment. Alec Jeffrey's DNA fingerprinting technique has brought justice to large numbers of people, not least to those denied entry to Britain because they did not have the resources or the evidence to demonstrate beyond doubt their relationship to relatives living here.

In a symposium whose purpose is to explore the interaction between genetics and society it is worth examining the public concern that genetics (and other scientific developments) engender, demonstrated by the extensive restrictive legislation surrounding DNA technology and by the title of Baroness Warnock's contribution, *The Problems of Knowledge.*

Obviously knowledge can be exploited in undesirable ways, and its application may have unexpected and deleterious side-effects. Yet, on any rational scale of values, the record surely shows that the balance has been overwhelmingly positive, and that there is an immeasurably greater problem – that of ignorance.

It is also important not to confuse 'public concern', which emanates from those who inform public opinion, with the idea that people at large fear and mistrust science and the technology it engenders. The public seems to me to take to science and technology like ducks to water – even dangerous technological developments like motor cars. It is perfectly capable of evaluating risks and rewards and balancing one against the other. The evidence shows that people are prepared to accept a substantial measure of risk, if they know what the risks are, in return for tangible rewards.

What people are negative about stems from the feeling that they are threatened by things outside their control – the belief that they are subject to risks they are not in command of and that they have no means of evaluating, without any tangible measure of personal reward. These are sensible fears, based as much as anything on a mistrust of the degree to which the expertise and objectivity of institutions (particularly commercial institutions) can be relied upon. The very words 'genetic engineering' engender just such fears. They give people the impression that they are to be the passive and unwilling targets of experiment.

How are people to be made better informed and therefore more confident?

□ The DNA scare

Public concern about the hazards of recombinant DNA technology was sparked by a letter to the journal *Science* (Berg *et al.*, 1974), signed by a number of distinguished molecular biologists, which advanced the view that recombinant DNA molecules constructed from different species were potentially hazardous to man and should be subject to a moratorium until the hazards had been properly evaluated. (Those who wish to follow the consequences of the *Science* letter can do no better than consult *The DNA Story* by Watson and Tooze (1981).)

The concern generated by this letter led to the establishment of advisory bodies in many countries. The main issue they had to address was the possibility that a laboratory construct might prove not only to be harmful in novel ways that could not be controlled, but might also spread in an epidemic manner. Their advice, and the subsequent restrictive legislation it led to, was designed to reassure an alarmed public. It had the opposite effect. A public with no other basis on which to form a view not surprisingly assumed that it must indeed be faced by real hazards if the issue was serious enough to attract so much attention from so many distinguished people and governments.

One conjectural hazard cited in the *Science* letter was the construction of bacterial plasmids (mini-chromosomes) modified to carry determinants, which would make these bacteria antibiotic resistant. These then might transfer such plasmids to bacterial strains that had not previously carried them. It is surprising that this should be seen as a significant hazard. The use of antibiotics in hospitals led long ago to the widespread natural evolution of such plasmids, which had picked up antibiotic resistance determinants from unknown sources. It is characteristic of these plasmids that as soon as exposure to antibiotic ceases they rapidly decline in numbers. In half a century of use I do not know of a single example of such a plasmid becoming widespread except in an antibiotic-contaminated environment. The only plausible interpretation of this is that a naturally sensitive bacterium pays for being antibiotic resistant, and that the price is too high for resistant strains to predominate in an antibiotic-free environment. Even if this were not the case, it is implausible to suppose that a pathogen that had acquired new properties as a result of recombinant DNA technology would not be as susceptible to the armoury of therapeutic techniques available to medicine as any existing pathogen.

It is significant that the many advisory bodies charged with assessing the potential risk of laboratory constructs did not clearly point this out. The public would have had no difficulty appreciating the analogy with domesticated plants and animals produced by more orthodox breeding methods. Every gardener knows that, if he takes a rest, it is the genetically unmanipulated weeds that take over and wipe out the genetically manipulated plants he treasures, not the other way round. I cannot recall a single recorded case in the history of agriculture of a domestic plant or animal being able to survive and spread in competition with wild organisms, just as bacteria resistant to antibiotics lose out in competition with their native progenitors.

There are of course many instances of the harmful effects of the transfer of organisms into new environments – for example, the fungus causing Dutch elm disease, which came from the United States to Britain – but in all cases the organism was a genetically *unmodified* wild-type.

Perhaps the members of these advisory bodies were unwilling to risk their reputations by taking an objective and optimistic view, given that all human endeavour carries risks. They were, in any case, given an impossible remit – to identify hazards without any theoretical basis on which to do so, and to try to quantify risks when the hazardous event frequency to date is zero.

It might be argued that advisory bodies with power to authorize or prohibit the construction or release of novel organisms will nevertheless protect the public from untoward events. Is this so?

If a novel organism proved to be harmful for completely unforeseen and unforeseeable reasons no advisory body would have prevented its release. On the other hand, an organism whose harmfulness was predictable might be released clandestinely and deliberately. An advisory body would not stop this either. The only accident that an advisory body could hope to prevent would be one that arose from ignorance or negligence. But do such bodies provide the best safeguard against such accidents? If their advice were to prove ill-founded they would not be personally liable, and individuals or institutions acting on their advice would have a perfect defence. Far better, it seems to me, to place explicit and unambiguous responsibility for such untoward events on those individually and corporately responsible. 'Depend upon it, Sir, when a man knows he is to be hanged in a fortnight it concentrates his mind wonderfully', as Dr Johnson observed.

This is not an argument against advisory bodies or advice. It is an argument about responsibility; about making sure that it lies where it clearly belongs and where it will be most effective. It is an argument for a permissive approach (with personal and corporate responsibility) rather than an authoritarian one.

No doubt it will be argued that this would inhibit research and development. Yet public corporations do not appear to be inhibited by the fact that they are being made increasingly responsible for faults in their products. Liability should arise from culpability, not from bad luck. In any case, bad luck will not be influenced by the existence of advisory bodies.

Of course legislative backup is important, but only where there are identified hazards that can be avoided by specific precautions. However, the legislation constraining the exploitation of recombinant DNA technology deals only with conjecture. Such legislation seems to me to be at best absurd and at worst a fraud. It is absurd if its purpose is to say 'We don't know whether this kind of experimentation is hazardous but, just in case it is, don't do it'. That is a recipe for ending all exploration and experimentation. It is a fraud, and a counterproductive one, if its purpose is to allay public concern or to insure the government and the scientific establishment against criticism should something untoward happen. Scientists who have participated in this fraud have only themselves to blame if their work is constrained as a consequence without their benefiting from a corresponding gain in public confidence.

□ Enhancing confidence

How, then, to address the antipathy (of the kind I have described) to science, scientists and technology? The need for education is often cited in this connection, but I have the impression that traditional modes of education are not very effective. People learn most easily from experience. They learn through their hands and by participation, and by analogy with things they already understand, rather than through formal instruction. They have an encouraging capacity to understand and come to sensible conclusions when they *need* to understand and have an incentive to do so.

One place to start in improving public knowledge of and confidence in biotechnology might be at the doctor–patient interface, since this is where people have their most direct contact with it. The patient is generally no more than that - a passive recipient of treatment - while the doctor plays the role of authority figure. Yet every patient is unique and every treatment is, or should be, an experiment. Why, then, do not doctors engage their clients (a much better word than patient, surely) on equal terms in the experiment of treatment to the benefit and education of both?

Many will have encountered in one form or another the experience I had when my ageing mother needed medication for high blood pressure. The doctor prescribed pills. She disliked the effect and took only half the amount he prescribed. He examined her later and pronounced himself well pleased with the result, while she said to me that he could not be much use if he did not even know the right dose!

Rather than this comical lack of communication, would it not have been more educative for both of them if he had given her a device to determine her own blood pressure and told her to discover for herself the minimum dose that worked? It is not a complicated procedure, after all. Most routine medicine is pretty simple. General practitioners therefore often seek to maintain their prestige and authority by mystifying it, following a long tradition practised by witch-doctors.

In this connection I was delighted to encounter the work of Dr Michael Rosen and his colleagues through an eloquent article by Professor Michael Rennie (1990). I will quote from it at length.

> Most people expect to have some pain after an operation, but too often they are not prepared for its intensity, especially after major surgery. The drugs needed to control this kind of pain are powerful, on controlled lists, and in other situations are associated with addiction. Many assume that they are the drugs least likely to be suitable for self-administration. [yet] pioneering work in Cardiff has resulted in a successful system for patient-controlled analgesia (PCA). One remarkable [*sic*] effect of this system is that when patients are in control they tend to give themselves a smaller dose than the hospital staff would administer.

> Since the drugs are controlled substances there is extra administrative work involved in getting the drugs released from cabinets and signing out medication. Patients, meanwhile, soon realize that asking for pain relief more often than nursing staff consider normal can lead to irritation and stress for themselves and the nurses. [Yet] variation in the speed at which drugs are absorbed [and also variation in body weight, presumably] means that pain relief can also be variable. Every patient needs to be treated individually.

The Cardiff team developed a pump connected to an intravenous tube and controlled by a push button that enabled patients to self-administer a pre-set amount of painkiller when needed, with a delay feature to prevent their administering too much. Patients were found to administer less than the normal clinical dose. Postoperative recovery was enhanced (perhaps because lower doses reduced the risk of drug interference with the important internal control systems that maintain oxygen blood level) and patients often left hospital earlier. Money was saved, and nursing staff were pleased because patients were more relaxed and less demanding.

When PCA was used in Australia on children as young as six the amount of drugs used was halved and the children were less distressed, making it easier for nursing staff to care for them.

One could scarcely find a more convincing illustration of the value of allowing people to be active participants in their own treatment rather than passive patients.

□ Ethical issues

Genetics and biotechnology face an additional burden from 'public concern'. They are liable to be seen as 'meddling with nature' or 'playing god'. Questions of safety are consequently compounded with ethical and moral questions. The recent discussion and legislation surrounding surrogacy and *in vitro* fertilization is a case in point (a surrogate mother being someone who bears a child on behalf of another person or persons).

In Britain the emphasis of current legislation is on prohibiting all commercial involvement in surrogacy arrangements. It thereby ensures that those seeking advice about surrogacy are effectively denied it except from their medical advisors (who may be paid, of course). Yet the key questions surrounding surrogacy are not medical.

They are moral, and therefore surely the responsibility of the participants, and even more clearly legal (what, for example, are the rights and responsibilities of the parties involved in relation to the child), yet are largely ignored by legislation and are questions which medical advisors are unqualified to address. Thus, the legislation makes patients out of healthy people and removes from them the ability to make critical decisions, based on informed advice, about their own lives.

Discussion of the responsibilities of participants in surrogacy has a long history. Thirty years ago Herman Muller, who discovered that X-rays cause mutations, argued that in the other kind of surrogacy – artificial insemination – it was wrong for doctors to be the arbiters of destiny in the choice of the sperm donor (Muller, 1961). He called for the establishment of banks of sperm from people with exceptional abilities of all kinds, which infertile families (or families at risk of having children with severe inherited abnormalities and therefore wishing to refrain from having children of their own) could draw upon after the donor was dead. It was a permissive notion, not an élitist one. People would be free to choose a donor who matched the physical and mental attributes they themselves desired and thus improve the odds of having a child with similar qualities. Placing the responsibility of choice in their hands would cause people to think responsibly about the issues, to take advice, and learn what these issues were. To which I would add that it would ensure that the child knew who it was.

Muller called what he was advocating 'germ cell choice'. He was concerned, not only with improving the human genetic endowment, but with the positive consequences of parents being in charge of, and responsible for, their own destiny rather than their having such profound decisions made by someone else on their behalf. This would create a better informed and more critical public perception of the issues.

☐ Conclusion

The point I am trying to make in this contribution is that, if we wish people to feel confident that biotechnological developments are a boon rather than a threat, we must enable them, and give them an incentive, to be active rather than passive in their exploitation. This will only occur in a permissive and participatory society.

Unfortunately, those who determine policy in a country like Britain often give the impression of being driven by authoritarian attitudes. The damage such attitudes can do is well illustrated by the failure of the British government's approach to limiting the spread of HIV infection. Controlling AIDS requires preventive action to be taken when the incidence of the disease is low. But this is just the period when the perceived risk of infection is too low to outweigh the rewards of a sexual encounter. In an anonymous article in the *Independent* newspaper (1990) an AIDS education worker, no less, confessed how, despite his involvement in this work, he and his female co-workers frequently engaged in sexual encounters without protection. He was behaving in a way that is familiar to all of us. We have a very high threshold of response to risk. If we did not, the sum of all the small risks we face every day would prevent our ever getting out of bed. This is also why even those who regard themselves as responsible citizens still drink and drive. Taking a chance is a pragmatic response to living in a risk-filled world.

This high risk threshold ensures that a promotion campaign that concentrates on warning people about the dangers of unprotected sexual intercourse will surely fail until it is too late – until the risks have become tangible, as they have for homosexual men and drug injectors.

To have any chance of success a campaign would need to focus on what is perceived as a near certainty – that if one is HIV-positive and engages in risky activity one will almost certainly harm one's partner. In other words, what is required (Pritchard, 1989) is a 'get tested' campaign directed at those who engage in risky activities. The theme might be 'don't kill your partner – get tested'.

Such a strategy, based on the assumption that most people have a responsible concern for the safety of others, would totally alter the equation. It would create a high risk for reckless behaviour towards one's partner and a high reward for being tested. It is another example of the value of knowledge in the right hands. Such a strategy would allow people to make informed decisions, and therefore responsible decisions, rather than decisions justified by ignorance. It presupposes that informed people are more likely to be responsible than irresponsible.

Anti-AIDS publicity in Britain assumes the opposite. Supporting a proposal that there should be blind and random testing of pregnant women for HIV (with the result not available to them), the Chief

Medical Officer of Health was quoted in the *Guardian*, on 4 November 1988 as saying 'AIDS specialists believe there is no particular advantage to the individual in knowing if he or she is infected since there is no effective treatment'. This is to look at the problem from a narrow, treatment-oriented point of view, rejecting the possibility that knowledge is more powerful in influencing personal behaviour than preaching.

All my experience suggests that, if people are given responsibility and equipped with the knowledge to enable them to make informed decisions, most will behave responsibly. They will certainly behave as responsibly as those who think it their job to determine what others should think and do because they are too frail and selfish to do so for themselves.

The perspective of these introductory comments may be different from that found elsewhere in this collection of essays. It is a personal view, which readers may find useful as they read the following essays.

References

Anonymous (1990). Confessions of an AIDS prevention worker. *Independent* (London) 27 November

Berg P., Baltimore D., Cohen S.N., Davis R.W., Hogness D.S., Nathans D., Roblins R., Watson J.D., Weismann S. and Zinder N.D. (1974). Potential hazards of recombinant DNA molecules. *Science*, **185**, 303

Muller H.J. (1961). Human evolution by voluntary choice of germ plasm. *Science*, **134**, 643–9

Pritchard R.H. (1989). Certainties that we can't ignore. *Guardian* (London) 20 March 1989

Rennie M. (1990). I'm the only one who knows how much it hurts. *Independent* (London) **1263**

Watson J.D. and Tooze J. (1981). *The DNA Story. A Documentary History of Gene Cloning* San Francisco: W.H. Freeman

2

The Human Genome Initiative

Dr James Watson

Jim Watson shared the Nobel Prize with Francis Crick and Maurice Wilkins for elucidating the structure of DNA, 'the molecule of life', at the Cavendish Laboratory in Cambridge in the fifties. He is now Director of the Cold Spring Harbor Laboratory on Long Island, in New York State, and was until recently coordinating the Human Genome Initiative.

The aim of this massive and controversial research effort is to sequence the four-letter alphabet of the chemicals *Adenine, Guanine, Cytosine* and *Thymine* (AGCT), which are represented 3 000 000 000 times in the human being, and make up the complement of human genes (see Box 1 on the unwinding of a DNA molecule). The sequence would fill the equivalent of 134 sets of the complete volumes of the Encyclopaedia Britannica, and will cost $2–3 000 000 000. The aim of this huge undertaking is to record the complete catalogue of human genes, in order to provide a reference for doctors and scientists who wish to study the genetic basis of inherited and acquired diseases, human development and evolution.

The Human Genome Initiative is a worldwide research effort that has the goal of analysing the structure of human DNA and determining the location of the estimated 100 000 human genes. In parallel with this effort, the DNA of a set of model organisms will be studied to provide the comparative information necessary for understanding the functioning of the human genome.

The information generated by the Human Genome Initiative is expected to be the source book for biomedical science in the 21st century and will be of immense benefit to the field of medicine. It will help us to understand and eventually treat many of the more than 4000 genetic diseases that afflict mankind, as well as the many multifactorial diseases in which genetic predisposition plays an important role.

US participation in the Human Genome Initiative is being coordinated primarily by two federal agencies, the National Institutes of Health and the Department of Energy. A joint five-year plan for the US Human Genome Project recently was prepared by the two agencies. The plan outlines the specific scientific goals to be achieved in the first five years of the US project, together with the rationale for each goal. Five-year goals have been indentified for the following areas, which together encompass the US Human Genome Project.

- Mapping and sequencing the human genome
- Mapping and sequencing the genomes of model organisms
- Data collection and distribution
- Ethical and legal considerations
- Research training
- Technology development
- Technology transfer

The five-year goals that have been specified in this plan assume the programme will rapidly reach a funding level of $200 million per year. This level of funding has been recommended by several groups of advisors. It is estimated that the project will take about 15 years to complete at that rate of funding. The plan will be reviewed annually and updated as further advances in the underlying technology occur.

The objective of the Human Genome Project is to work out the complete genetic constitution of the human being. There are three billion pairs of letters in human DNA (see Box 1 on unwinding a double helix of DNA) and, according to conventional wisdom, about 100 000 genes. I will define a gene simply as 'the DNA unit that first gives rise to an RNA copy (messenger) which is then translated into a specific protein, the molecules that determine the outward characteristics of all cells and organisms. I think that eventually we will find closer to 200 000 or 300 000 genes in the human blueprint, because I suspect large numbers of small genes will be found. Only time will tell. I also suspect the number of known fruit fly (*Drosophila*) genes will increase to at least 20 000, and I just cannot accept that the fruit fly might have the same order of magnitude of genes as we do! My chief aim was to help develop an American programme, but we know the human genome is too interesting to be left just to the United States. Many other countries, including the UK and, in particular, France and Japan, are developing big programmes.

□ Origins of the Human Genome Project

There were several origins of the Human Genome Project in the United States. The first was a proposal to do the whole thing at the University of California in Santa Cruz (UCSC). This proposal had an interesting initial objective. UCSC was going to build a very big telescope, but eventually they lost it to the California Institute of Technology. They worried about which big project could replace their telescope and so, in May 1985, Bob Sinsheimer, then Chancellor at UCSC, held a small meeting about working out the human genome. Among the attendees were Wally Gilbert, Charlie Cantor and Lee Hood, all of whom are well-known figures in American biology and were enthusiastic to get the project started.

The second proposal for a genome project was put forth by the Department of Energy (DOE) in the spring of 1986. The project appealed to them because they needed DNA sequence data to monitor mutation rates, and they know how to handle high-technology programmes. In particular, the DOE national laboratories, which have been so successful in high-energy physics research, believed they had

Box 1
Unwinding of a double helix of DNA

A double-stranded DNA molecule unwinding and replicating. The four-letter alphabet A,G,C,T represents adenine, guanine, cytosine and thymine, the four nucleotide bases. Guanine pairs with cytosine and thymine with adenine, and hydrogen bonds between the base pairs provide stability to the molecule. Replication occurs when the parental strands unwind, after which nucleotides in the 'cellular soup' pair up with the exposed bases of the parental strands. The human genome has 3 000 000 000 base pairs, and one of the aims of the Human Genome Initiative is to 'read' the complete human DNA sequence. Figure reprinted from Strickberger M. (1976). *Genetics* 2nd edn. New York: Macmillan and based on Stent G.S. (1971). *Molecular Genetics: An Introductory Narrative*. San Francisco: W.H. Freeman.

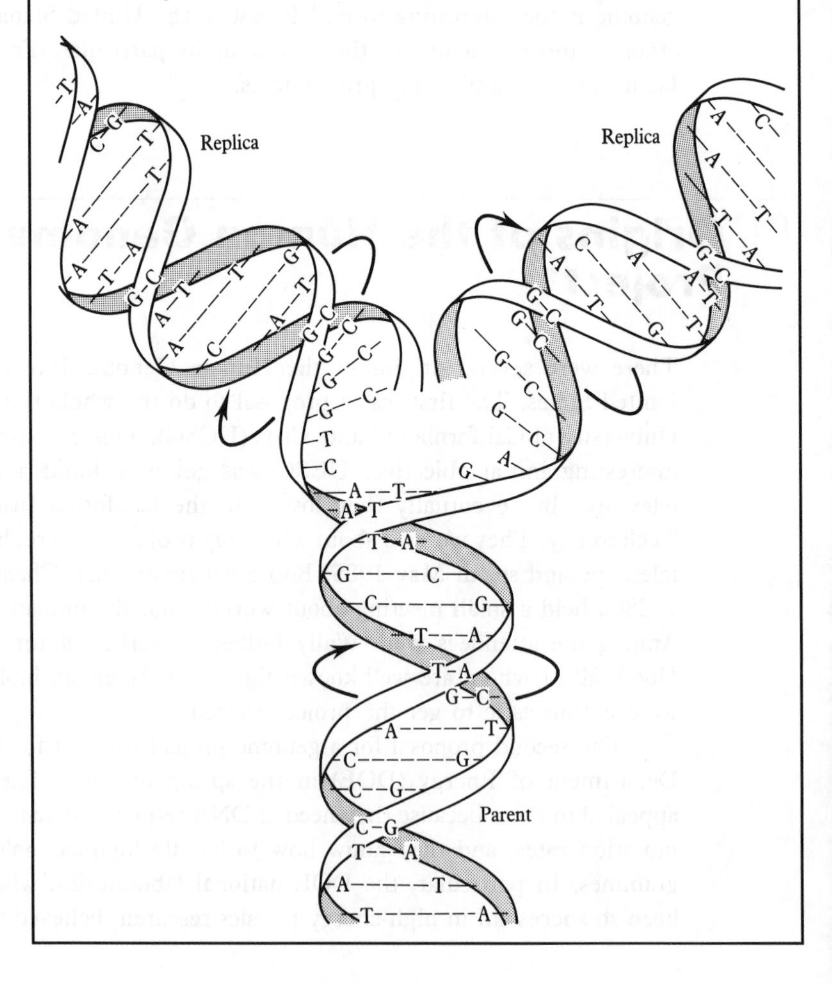

the brains and the technological skill to sequence the human genome. A number of us took the DOE proposal quite seriously, because who could really be against sequencing the human genome? Clearly, we all want to know the exact genetic blueprints for human existence.

But many of us felt that the agency dominating this big American programme should be the National Institutes of Health (NIH) since it already funded most people interested in DNA through a rigorous peer-review process. Thus in the spring of 1988, David Baltimore and I helped persuade our Congress to set up a second human genome programme at the NIH. We did not try to block DOE genome funding, arguing that two efforts might engender healthy as opposed to unhealthy competition. As long as we did not do exactly the same thing everything should be fine! So in 1990 we had two major agencies funding genome research in the United States with the amount of money going to NIH being roughly twice that which goes to DOE. Happily we worked closely together, and there was very little friction between us. Therefore, I was appointed Director of the division within NIH called the National Center for Human Genome Research (NCHGR). In Bethesda, Maryland, I had an initial staff of about 40 people who were there to reassure the biological world that the money was being spent on the basis of peer reviews and that the grant money would be distributed wisely.

□ Financing the initiative

In the fiscal year 1990 (FY90), we had almost $60 million to spend on the NIH genome programme. The DOE figure was a little under half that. In the budget sent to Congress by the White House for FY91, our programme was to receive $108 million and DOE $48 million, with the money coming out of the Office of Management of the Budget (OMB). In 1990, reflecting the healthy debate on the genome project, a number of molecular biologists wrote letters to Congress saying that our programme was already too large and was hurting their own research programmes by competing for scarce NIH funds, and that our budget should not be increased at all. My hope was that supporters of our work would also write letters to Congress, pointing out the importance of our programme for all forms of medical research. (The operating budget for the NCHGR in FY91 was finally set at about $87 million.)

How did we go about spending our initial money? To start with, largely on ordinary research grants as opposed to production-type mapping and sequencing projects. The main point of these initial grants was to develop ways by which mapping and sequencing could be done more cheaply. All the early objections to the Human Genome Project focused on its cost. On the basis of technology that existed at the time, it would have cost about $30 billion, not the $3 billion which we initially told Congress should be put aside for the project budget†. Of course, all this money would not be spent in one year. Our best estimate today is that some 15 years will be required to complete the work. When I sat on the Human Genome Committee set up by the US National Academy of Sciences in 1986, we said the project should initially devote itself to finding ways to reduce costs. In 1990, at the very best laboratories, sequencing could be done for $3 per base pair, but in many laboratories more than twice this sum was required. Often much lower figures were reported, but these estimates disregard many key components, such as the costs to prepare the DNA for subsequent sequencing, to build the laboratory building or to pay for the janitors and utilities. When all the true costs are added up, I think most current (1990) sequencing efforts required $5 per base pair, and that was an unacceptable cost for megabase sequencing efforts. We knew we had to reduce the cost by a factor of ten if we were to complete the programme‡.

We estimated that the sum of $3 billion should fund 30 000 person–years of labour. For a 15-year project, we expected roughly 2000 people to participate. Some scientists said there was no reason to do it over 15 years. Why not do it over 25? One important reason is that if you did it over 25 years, most of the experienced scientists involved in it might be dead, at least mentally, by the time it was finished. It is best to have a programme where those who plan it will actually see its fruition. If you were building an accelerator and said it would not be finished until the people who planned it were dead, you would wonder about their enthusiasm. Most people like to do things where they can see the results.

† *Editors' note* Currently (1993) the contribution of NIH alone has so far been over a billion dollars.
‡ Costs of sequencing have fallen dramatically in recent years, and in many cases a cost of 50 cents per base is achievable, although certain sequences can prove much more difficult and therefore more expensive to elucidate. With this and other technical developments, the pace of advance is accelerating markedly, and 15 years to achieve most of what we want from the program may be an overestimate.

Of course, if the project were planned by much younger scientists, you could afford to do it over a much longer interval. But I find that the older people are, the more enthusiastic they are for the project. This difference between younger and older scientists is easy to explain. The older you are, the more likely you are to think about disease. When you ask a supporter of the project 'Why do you want to do it?' the overriding reason is that it will let us get handles on genetic disease much more efficiently. There are, of course, many excellent younger scientists who are keen for the genome programme, the majority for its intrinsic benefits to pure science. In other cases, their interest primarily reflects serious genetic disease within their families or among their friends.

□ Knowledge of the genome will aid the fight against disease

We have had enormous success in understanding the fundamental molecular basis of cancer over the past 10 or so years, and all this progress has come from finding first oncogenes and now anti-oncogenes. If we had not been able to study cancer at the DNA level, it would be hopeless. Masses of research money would continue to go to the National Cancer Institute or the Imperial Cancer Research Fund, and there would be very little coming out. But now money can be spent on cancer very intelligently.

Equally important, genetics research can have the same impact on a number of other diseases. We still have no idea, for example, what causes Alzheimer's disease, except for the fact that in some families there appears to be a genetic predisposition, especially in cases of early onset. Getting the gene(s), however, may be very difficult, in part because family histories with accurate diagnoses are hard to get. Many of those with the gene will be missed, since they will die of another disease before they come down with Alzheimer's disease. To actually get the gene, we may need to know the sequence of quite a significant region of chromosome 21, the chromosome to which many cases of early-onset Alzheimer's disease seem to map.† We may thus have to

† *Editors' note* Recent studies confirm that a gene on chromosome 21 is responsible for a minority of Alzheimer's cases. This is an important breakthrough, but much remains to be done.

develop megabase DNA-sequencing capabilities, if we want to get either at this horrid disease or any other where exact genetic mapping will be difficult.

You may have read that for a brief while the gene for manic depression within the Amish sect was thought to be on chromosome 11, but then several newly affected individuals were found who seemed to rule out this map assignment. In part, the current difficulty in finding this gene may lie in misdiagnosis. Also hindering the search, however, is the limited number of mapped genetic markers now available. We need more genetic markers accurately located along each of the 24 different human chromosomes (22 autosomes and the sex chromosomes X and Y).

Good genetic maps of humans were essentially impossible even to contemplate until the idea of using DNA polymorphisms as genetic markers (RFLPs – restriction fragment length polymorphisms) came along. This concept was first proposed in the late 1970s and by 1985 extensive genetic maps of most human chromosomes had become available (see Box 2 on DNA polymorphism). By 1990 about 350 of these DNA polymorphisms had been put on maps. At first this was reassuring, and it was widely thought that we had enough DNA markers to map the genes behind most genetic disorders to the chromosomes on which they reside. But at a meeting in March 1990 in Salt Lake City, Ray White and Helen Donis-Keller, who had previously mapped polymorphic markers, said that they really had only about 150 highly heterozygous markers and at least twice this number was required to make a good map.†

So we made the decision to have a crash programme to put about 300 highly informative markers on the genetic linkage map. In the process, we decided to pass out money to about ten major American laboratories to fill in the gaps, with the intention of making them available to anyone in the world who wanted them. Sending out to the human genetics community a large number of these probes was a first major achievement of the NCHGR. Now I feel slightly stupid that in 1988 we did not initiate this crash genetic map effort. Conceivably, we would not now have the embarrassment that the map position of the gene for manic-depressive illness within the Amish is still unknown.

† *Editors' note* This problem has now been overcome: with new methods of analysing DNA sequences, thousands of such polymorphisms (representing different genes) have been mapped. Such mapping whilst telling us nothing of the nature (sequence) of a particular gene, makes it infinitely easier for subsequent isolation and sequencing.

Box 2
DNA polymorphism

A defective gene causing a specific human disease is located near a region of DNA (a marker) that has a difference in sequence (it is polymorphic-RFLP) compared to its homologous sequence on the opposite chromosome. During meiosis in the mother, recombination between the two chromosomes produces eggs containing, more often than not, a single chromosome either that has both the disease locus and the marker or neither. Thus the disease gene and marker are closely linked spatially. Had the marker been located farther away on that chromosome, recombination would often separate the disease gene and the marker, that is, linkage would be absent. By taking several hundred polymorphic DNA markers and mapping them with respect to each other, we can generate a large-scale map of the human genome.

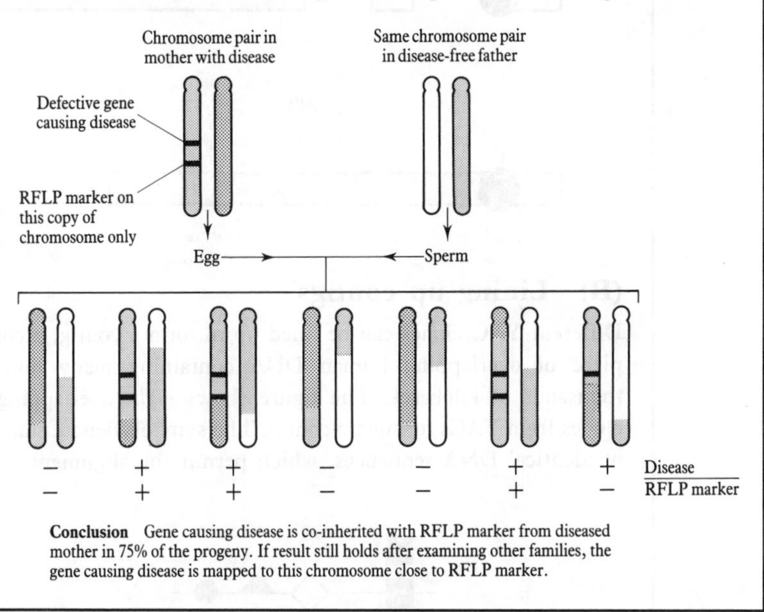

Conclusion Gene causing disease is co-inherited with RFLP marker from diseased mother in 75% of the progeny. If result still holds after examining other families, the gene causing disease is mapped to this chromosome close to RFLP marker.

☐ Practicalities of getting the sequence of the human genome

Our second major objective was to obtain a set of overlapping DNA fragments for all the human chromosomes (overlapping DNA libraries). Initially, the DNA pieces we were trying to overlap were

Box 3

(A) Yeast artificial chromosomes (YACs)

Cloning human DNA in yeast artificial chromosomes (YACs). The left arm of the yeast chromosome (L) plus the centromere (C) are ligated to a large piece of human DNA containing several hundred thousand base pairs of DNA, and the right arm (R) of the yeast chromosome. The YAC can replicate itself within yeast cells. Cosmids, which are small circular pieces of DNA, may also have human DNA fragments ligated into them, but only up to 50 000 base pairs (50 kb) can be incorporated.

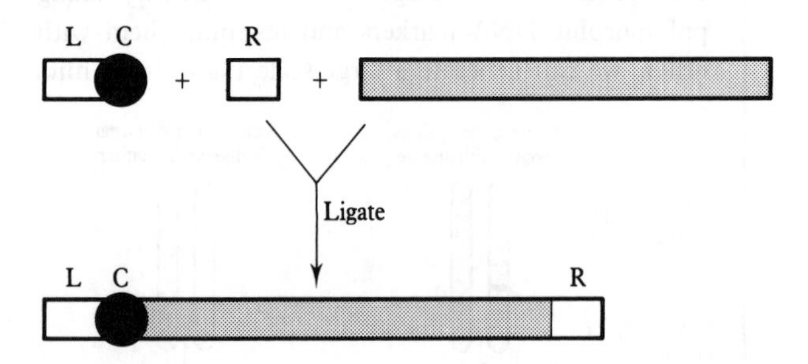

(B) Lining up contigs

Different YAC clones can be lined up to form a contig, a continuous piece of overlapping human DNA containing many hundreds or thousands of kilobases. The figure shows eight overlapping human clones from YACs forming a contig. The symbols denote small regions of identical DNA sequences, which permit the alignment.

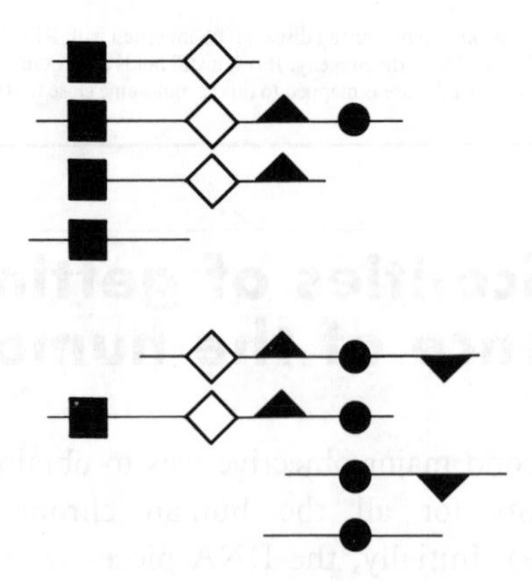

50 000 base pair fragments that had been cloned into cosmid vectors. Overlaps were looked for by finding common restriction sites, a method successfully employed by John Sulston and Alan Coulson at the laboratory of molecular biology in Cambridge, using nematode (*Caenorhabdytis elegans*) DNA. The longest human cosmid contigs (continuous pieces of sequence), however, that have been put together for human chromosomes seldom exceed 30 000 base pairs – that is, one thousandth of a genome – and their exact order along their respective chromosomes remains to be worked out (April 1990).

Fortunately, by the time we asked Congress to give us money for a genome programme, we knew we would not have to depend on cosmid overlapping. Maynard Olson at Washington University in St Louis had been developing a technique for cloning much larger pieces of DNA by making them parts of yeast artificial chromosomes, called YACs (see Box 3 on yeast artificial chromosomes). Initially, his hope was that he might be able to clone into YACs pieces of DNA roughly the same size as a typical baker's yeast *Saccharomyces cerevisiae* chromosome (about one million base pairs). This would mean that if you took the smallest human chromosome, 21, you would need only 50 YACs laid end to end together. Ordering such YACs would be fairly simple. In reality, the vast majority of YACs made in St Louis contain only 300 000–400 000 base pair inserts. Nevertheless, this was a factor-of-ten improvement over cosmids. The library in St Louis in 1990 had about 60 000 members, representing about a five-fold coverage of the human genome. So most human genes would be present in this library.

With such a YAC library and the DNA probe for your favourite gene, you could screen the library and get your gene out as a big continuous piece of DNA. Lots of people sent their probes to St Louis, saying 'Would you give me the YAC that covers my gene?' They rapidly had many more such requests than they could handle. About half the YACs they were initially screening were just for their colleagues in St Louis. In order not to be overwhelmed with screening for the rest of the world, the St Louis group passed out the library, and a companion library from Paris has similarly been expected to be passed out, too.† By 1990 in St Louis, they had already made two megabase overlaps of YACs in the chromosome regions that code for the gene for Factor 8, the necessary clotting factor lacking in many haemophiliacs.

We were thus very confident that it would be possible to clone all the human DNA into overlapping DNA fragments using the YAC

† *Editors' note* Material from the Paris group is now widely available.

technology. This optimism was expressed in April 1990 in Bethesda where most of the people in the world working on chromosome 21 came together and presented data on their various chromosome 21 DNA probes. It was agreed that the YAC library from St Louis would be sent to Denver, where David Patterson would screen everyone's probes against Maynard's library. Then we would have the YACs corresponding to the 35 or so good genetically mapped probes along this smallest of all human chromosomes. It was my hope in 1990 that soon we would have a complete overlapping YAC map for chromosome 21. (This was indeed achieved in 1992 – *editors*.) Obviously, there will be further technological improvements that will allow physical mapping to be done faster and cheaper. It is a question now of organization and finding the people who want to do it.

Once you have the DNA as overlapping YACs, the question is always one of how to get this DNA down to a form in which you can sequence it. So the major concern of the Human Genome Project is whether sequencing technologies will be developed enabling workers to go from YACs down to getting out base pairs at a cost of about $0.50 per base pair. (This is now largely the case – *editors*.)

The hope of most of us is that sequencing can be done by machines as opposed to the hands of scientists. At a meeting held in late 1989 outside Washington, three different laboratories said 'Yes, we finally can sequence big stretches of DNA from machines'. But this was not a widely held view among molecular biologists. The conventional wisdom in the United States in April 1990 was that the machines would not work because they were not accurate enough with the computer programs that read their gels, and made mistakes about one in 100 times. The machines' proponents countered that the software had been improved so that, in fact, the mistake level was down to about one in 1000. Again, only time will tell. At an early genome project peer review, the study section word was 'Don't give large sums of money to those who say that they can sequence by machine'. We should remember, however, that in 1988 the mood among peer reviewers was that YACs would not work.

The NCHGR in 1990 expected to fund a number of laboratories that wanted to take on megabase sequence projects and while doing so to develop new procedures that would drastically reduce the cost of sequencing. We did this knowing that some scientists would object, arguing 'Don't give out the money unless we know they can reduce the cost'. My answer was that most research grants do not lead to big breakthroughs, so why worry that some grants will not achieve their objectives? That is just the nature of science. But some people get very nervous at the thought that you might actually waste $10 million on a machine-sequencing effort that fails. It did not worry me as long as one low-cost sequencing procedure emerges. By 1995, if we actually have a method that can reduce the cost to $0.50 per base pair, I will not be tarred and feathered by the American molecular biology community.

□ Ethics and the genome project

The last area of our programme that I wish to mention is ethics. When we are able to look at the individual genes of people, we are going to discover many imperfect ones. How can we block the release of facts about our individual imperfect genes to insurance companies, rival political parties, and so on down the line? At the level of the US Congress, many are worried about whether the Human Genome Programme will make us able to identify the genetic undesirables and make their lives even worse. I believe that we shall have to have laws stating very strictly that the only person to have the right to authorize looking at, say, my own DNA should be myself, except in cases of criminal behaviour. I think no one should have the right to screen another's genome to see if he or she is missing an anti-oncogene or going to be susceptible to juvenile diabetes. I am optimistic that eventually laws will be passed that say 'You can't do it'. To block the whole genome programme because of fears that it might be misused would be to say that we should not really try to find the gene for Alzheimer's disease, because if we knew where the gene was, we might be able to screen people and say 'Ah, you're going to become senile at 55, and so we're not going to employ you or let you have insurance'. So what is worse: not finding the gene or worrying about whether it will be misused? I think that by far the worst thing is not to find the gene. We should first find the gene, and then ensure that it will not be misused at the public level. When I testified before a Congress committee in 1990, a thoughtful member responded with his belief that our programme should not have its budget further increased until we had solved the resulting ethical dilemmas. I asked him 'Should we stop work on putative Alzheimer's genes until we're on top of the resulting insurance and employment dilemmas?'.

It reminds me of some of the discussion that surrounded recombinant DNA technology at the time of the Asilomar Conference in 1975.

I was tempted then to put together a book called the *Whole Risk Catalogue*. It would contain risks for old people and young people and so on. It would be a very popular book in our semi-paranoid society. Under 'D' I would put dynamite, dogs, doctors, dieldrin and DNA. I must confess to being more frightened of dogs. But everyone has their own things to worry about.

When I accepted my position in September 1988, I proposed that we should spend three per cent of the genome budget, which would soon mean some $6 million, to get the various ethical, legal and social issues raised by the Human Genome Project in front of public consciousness. Some of my friends worried that our talking about ethics would backfire and unnecessarily heighten human fears about the misuse of genetic knowledge. But I know there will be serious ethical dilemmas for which we will have to be prepared. So besides thinking through how we're going to get the human genome sequence, those of us who have been involved in leading the effort have had to spend a large fraction of our time talking to the public, explaining why we are doing it and why we think we are good guys and not bad guys. It is not always smooth going, but neither is life itself.

3

Molecular Medicine: Diagnosis and Therapy

Dr C. Thomas Caskey

Thomas Caskey is Director of the Institute for Molecular Genetics at the Baylor College of Medicine in Houston, Texas. He is an advisor to the Human Genome Project, and his own work has focused on the molecular genetics of inherited human diseases. He has pioneered molecular analyses of such diseases as Lesch–Nyhan syndrome, which is associated with self-mutilation, and has developed revolutionary techniques of examining the mutations causing inherited diseases.

The X chromosome offers a special diagnostic challenge for the identification and diagnosis of its some 400 heritable disorders, since the severe disease loci are prone to new mutations in succeeding generations. We have developed molecular methods that can scan for unique mutations at specific loci such as Lesch–Nyhan syndrome, steroid sulfatase deficiency, ornithine transcarbamylase deficiency, and Duchenne muscular dystrophy. These techniques include automated DNA sequencing of mutant alleles, multiplex PCR scanning for deletions and duplications and mismatch cleavage of mutant/wild-type heteroduplexes. We have now developed a strategy for scanning for disease loci and candidate expressed sequences (that is the tell-tale messenger RNA made as a copy of a gene) previously uncharacterized from mapped yeast artificial chromosome clones of the human X chromosome. This new genome strategy is expected to reduce the cost and effort of new disease gene identification.

Therapeutic attempts at genetic correction are discussed for the X-linked ormithine transcarbamoylase deficiency and autosomal recessive adenosine deaminase deficiency. Each offers special biological and strategic problems that have been partially solved.

The interface between molecular genetics and medicine has generated much excitement in the light of recent developments. The field of medical genetics, with its feet in both arenas of genetics and medicine, has been in an excellent position to transfer technological advances into medical practice at a brisk pace. As recently as 1976, W.Y. Kan and colleagues were the first to apply the tools of recombinant DNA technology to the diagnosis of a heritable disease, and the application of this technology to the field of medical genetics has been expanding rapidly ever since.

□ The Human Genome Project

The community of medical geneticists in the United States has welcomed the concept of the Human Genome Project, a huge international collaboration to define the genetic make-up of man. In Chapter 2 Jim Watson has outlined the major features of this vast undertaking, and my aim is to illustrate its importance to the success of our field.

Isolation of X-linked disease genes

My particular area of investigation is the human X chromosome, which contains about 5% of the total genome, including many genes responsible for disease. Disease genes are easily mapped to the X chromosome because males carry only one X chromosome, and so-called 'X-linked' diseases typically affect males only. Females carrying two X chromosomes are usually protected from these disorders, although they may act as carriers of the diseases. Figure 1 shows a diagram of a human X chromosome with some examples of X-linked disorders found within it; some have been assigned to a specific chromosomal location, others remain unmapped so far. It is currently estimated that there are 350 or more X-linked disorders in humans.

In the past, genes were identified and characterized by one or more research groups focusing on a particular disease entity, determining an approximate chromosomal location for the appropriate gene, and then investigating in detail that particular region of DNA. A recent example of this approach is the isolation of the X-linked Duchenne muscular dystrophy gene in 1987, which was a heroic

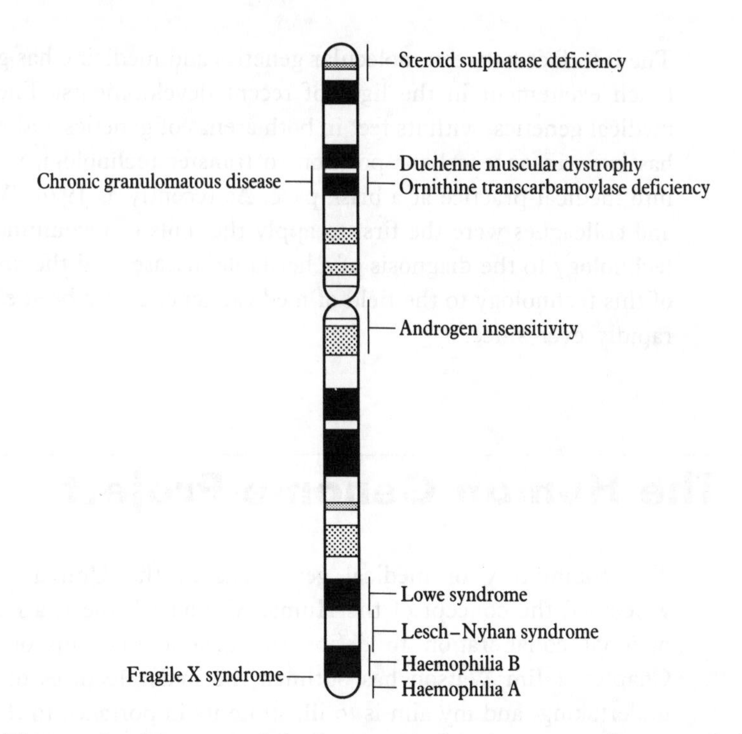

Figure 1 Diagram of the human X chromosome and some X-linked genetic diseases.

A human X chromosome divided into 'bands' (differently stained regions observable under the microscope, used to identify and subdivide each chromosome). Alongside are marked some of the disorders whose genes are located on the X chromosome. The larger the region of the chromosome identified with each disorder, the less precise is the current mapping information.

undertaking. The intensity of effort, the number of investigators and the cost involved indicate that this was an unreasonable approach to the isolation of disease genes in general. An entire research foundation supported the isolation of the Duchenne muscular dystrophy gene, but there is no foundation to support a similar effort for panhypopituitarism, or any one of the 5000 or more other heritable diseases in man. The 'genome approach' provides an alternative strategy.

To clone the entire X chromosome in yeast artificial chromosome (YAC) vectors would take about 2500 clones (allowing for some duplication to ensure complete coverage), and this set of clones would contain the necessary genetic material for the study of any X-linked gene. Furthermore, it is possible to line up these YAC clones in order along the X chromosome from tip to tip so that there is rapid access to DNA from the area of interest. (See Chapter 2, Box 3.)

Two examples of disease entities encoded by the X chromosome and now mapped very precisely by the genome approach are Lowe syndrome and Fragile X syndrome (X-linked mental retardation). Had previous technologies been used, large groups and a significant amount of time would have been required to identify these loci, but with the current approaches it has taken a limited number of people only a short time.

Isolation of genetic landmarks

An important part of the Human Genome Project is the isolation of frequently spaced 'markers' along each chromosome for the purpose of laying out a genetic map. The most useful kind of marker for the medical geneticist is one that shows some variation between individuals and yet is present in everybody at the same genetic location. Such a marker is termed 'polymorphic' to indicate its appearance in different forms. Markers with larger numbers of variants are termed 'highly polymorphic' and are more informative. Polymorphic markers are inherited in a predictable manner and, if a marker lies close to a disease gene (that is, is 'linked'), following its inheritance within a family can give valuable information about the inheritance of a disease gene, and thus who is likely to be affected by the disorder or who is a carrier. (See Chapter 2, Box 2.) These markers can also serve as guides for focusing on the particular region of DNA in order to isolate the nearby disease gene. Figure 2 shows an example of the inheritance of a polymorphism within a family.

□ Mutation detection

Once a disease gene has been isolated, the material is available for investigating mutations within it that result in disease. I will give three examples here of different ways in which these mutations can be detected, using three X-linked disorders as models.

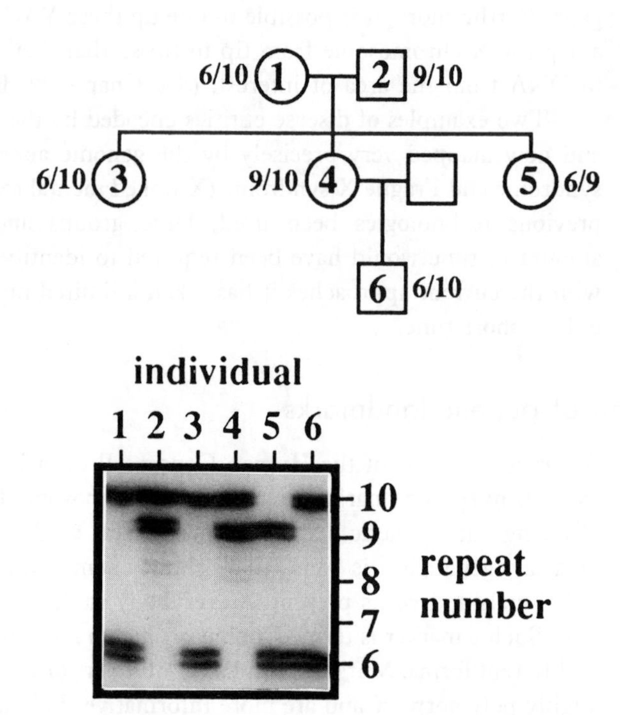

Figure 2 Example of polymorphism inheritance.

In the pedigree shown above, circles indicate females and squares indicate males; the numbers within the symbols relate to the individuals shown in the panel below. The analysis of individuals 1–6 indicates that each person has two markers, one inherited from each parent. The polymorphic markers in this instance are labelled 6–10 and the appropriate numbers are indicated beside each individual in the pedigree above.

(Source: Rossiter B.J.F. and Caskey C.T., 1991).

Duchenne muscular dystrophy

The most common lethal muscle disorder in males is Duchenne muscular dystrophy (DMD). After the gene responsible for this disease was cloned in 1987 it was found to have some interesting features. The gene is huge (over 2000 kilobases, whereas most genes range in size from 1–400 kilobases) and generates a large protein. Over half the mutations affecting this gene are deletions removing part or all of the gene and many of these deletions have an end in one of two

Box 1
The Southern blot

DNA is cleaved with several restriction enzymes: in effect, molecular scissors that can cut DNA. The short fragments of cut DNA are separated according to size by passing them through an agarose gel and applying an electric current. The gel is then laid on a nylon membrane and by the use of various buffers the DNA fragments are made to flow out of the gel and stick to the membrane. Thus the DNA fragments on the gel are transferred to the membrane and can now be hybridized to an appropriately labelled DNA probe. Any DNA fragment on the filter that has a similar sequence to the probe will fix the probe to the membrane and therefore produce a signal in an autoradiograph when the membrane is exposed to X-ray film.

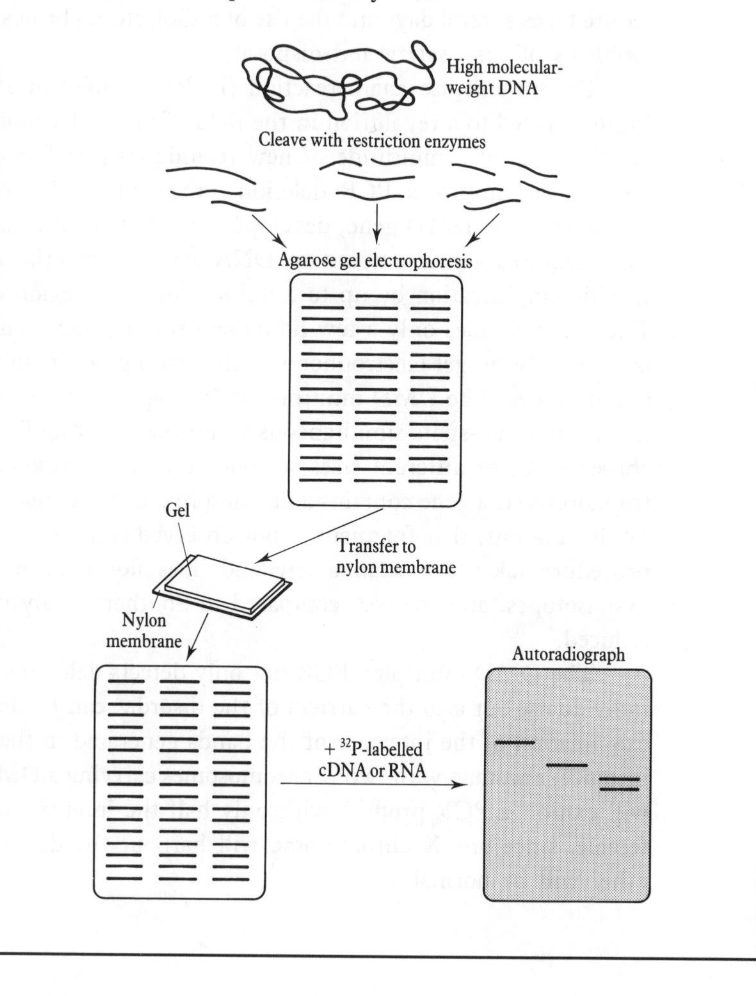

'hot-spot' regions. Although there appears to be a predisposition for a particular kind of mutation (deletion), and even a bias in the location of these deletions, each affected family actually carries a different mutation and thus DMD is defined as a 'new mutation disorder'.

Early methods of mutation detection within the DMD gene used Southern analysis (see Box 1) which involves the preparation of high-quality DNA from the X-chromosome, digestion of the DNA into smaller pieces with specific enzymes, size-separation of the resulting fragments by gel electrophoresis, transfer of the DNA fragments to a nylon membrane, incubation of the membrane with a radioactive probe and autoradiography to determine which of the fragments binds to the radioactive probe. An abnormal pattern compared to a normal sample indicates some sort of gene rearrangement. The whole procedure takes several days and the use of radioisotopes brings associated problems of cost, safety and disposal.

The polymerase chain reaction (PCR) invented in 1985 by K. Mullis has led to a revolution in the field of molecular biology, forming the basis of a multitude of new techniques (see Box 2). One of these is the multiplex PCR deletion screen for the investigation of deletions in the DMD gene, developed by J.S. Chamberlain in 1988. PCR requires a tiny amount of DNA starting material and allows specific amplification by up to a million-fold of a region of interest. The reaction takes only a few hours and the products can be visualized directly by gel electrophoresis, eliminating any requirement for radioisotopes. The DMD multiplex PCR amplifies nine regions from the deletion hot-spots simultaneously and the resulting fragments are chosen to be of different sizes so that they can be resolved by electrophoresis. If a gene contains a deletion in one of the regions covered by the analysis, that fragment is not observed (Figure 3). The entire procedure takes less than a day and does not require the use of radioisotopes, and the cost compared to Southern analysis is greatly reduced.

The DMD multiplex PCR not only detects deletions in affected individuals, but also the carriers of the disorder can be diagnosed by examination of the intensity of the bands generated in the PCR. For instance, a woman with two X chromosomes carrying a DMD deletion will exhibit a PCR product with only half the intensity of a normal female, since one X chromosome will harbour the deletion and the other will be normal.

Box 2
The polymerase chain reaction

Specific DNA sequences are amplified as follows. The double-stranded DNA is heated, causing the two strands to separate. Small DNA primers, A and B, have DNA sequences similar to part of the single-stranded DNA, then bind to the single-stranded DNA. By use of a heat-tolerant polymerase these primers allow the generation of two new chains complementary to the originals, giving two new double-stranded molecules. These are then heated to separate the strands, and the whole process is repeated 20 or 30 times. Each cycle gives a doubling of the DNA fragments that are being amplified. Thus small regions of DNA can be amplified from one molecule to 4, 8, 16, 32, 64 and so on. After 30 rounds of amplification a single molecule will be represented 2^{30} times. (Figure reproduced from Weatherall D.J., 1991)

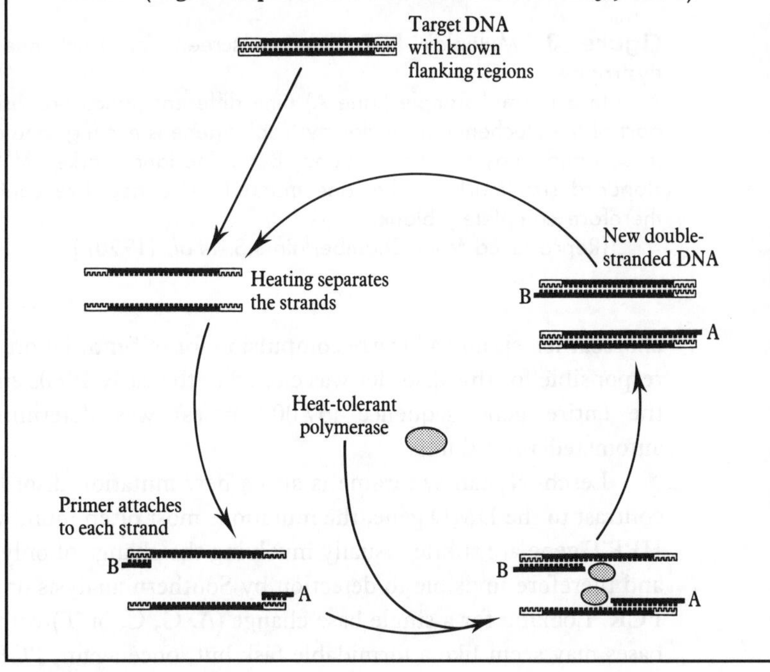

Lesch–Nyhan syndrome

The Lesch–Nyhan syndrome is a disorder of purine metabolism resulting from disruption of the hypoxanthine phosphoribosyltransferase (HPRT) gene. The symptoms include mental retardation,

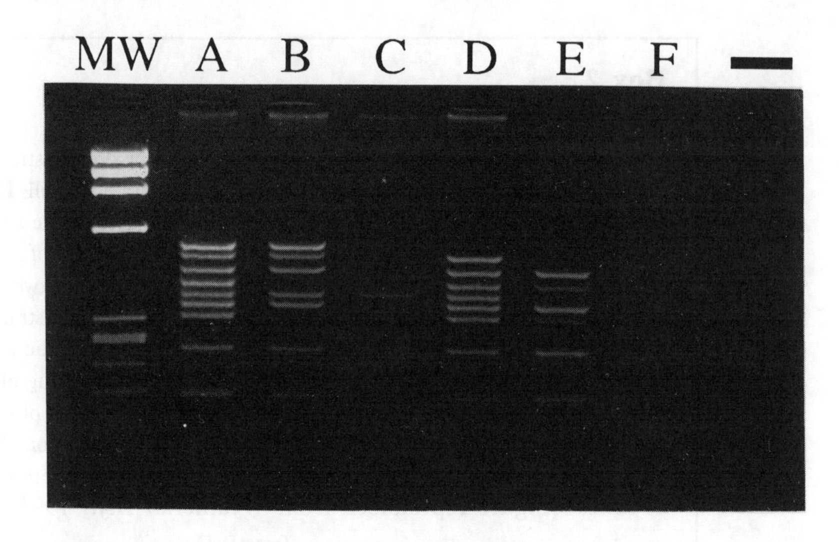

Figure 3 Multiplex PCR deletion screen for Duchenne muscular dystrophy.

In a normal sample (lane A) nine different bands are visible but, if part of the Duchenne muscular dystrophy gene is missing, one or more of these bands may be missing (lanes B–F). The lane marked MW contains standard size markers; the lane marked – is a negative control and is therefore completely blank.

(Reproduced from Chamberlain J.S. *et al.* (1990).)

choreoathetosis and a bizarre compulsion for self-mutilation. The gene responsible for this disorder was cloned in the early 1980s and in 1990 the entire gene sequence (57 000 bases) was determined using automated procedures.

Lesch–Nyhan syndrome is also a new mutation disorder but, in contrast to the DMD gene, the mutations most often found within the HPRT gene are subtle, usually involving the change of only one base, and therefore invisible to detection by Southern analysis or multiplex PCR. Looking for a single base change (A, G, C, or T) within 57 000 bases may seem like a formidable task but, once again, PCR has provided the means for such searches to be made efficiently. The first step is a multiplex PCR focusing on the regions of the gene that encode the protein final product; this reveals deletions in a few cases but the majority of samples from Lesch–Nyhan patients will appear normal at this stage. The main purpose of PCR is to supply material for the next phase which is automated sequencing of the PCR products; this can reveal single base changes very readily (see Box 3).

Box 3
DNA sequencing method developed by Fred Sanger

A dideoxynucleotide, either ddATP (adenine), ddCTP (cytosine), ddTTP (thymine) or ddGTP (guanine), is incorporated into a growing DNA strand, terminating chain growth. A primer is used to begin the chain formation; as shown within the circle, the ddGTP will stop chain growth at every position where the corresponding cytosine is found. Four different reactions are run, each with a different dideoxynucleotide. Products of each reaction are a series of incompletely elongated segments that can be separated by gel electrophoresis. The sequence can then be read from the bands produced in the gel, either by eye or automatically.

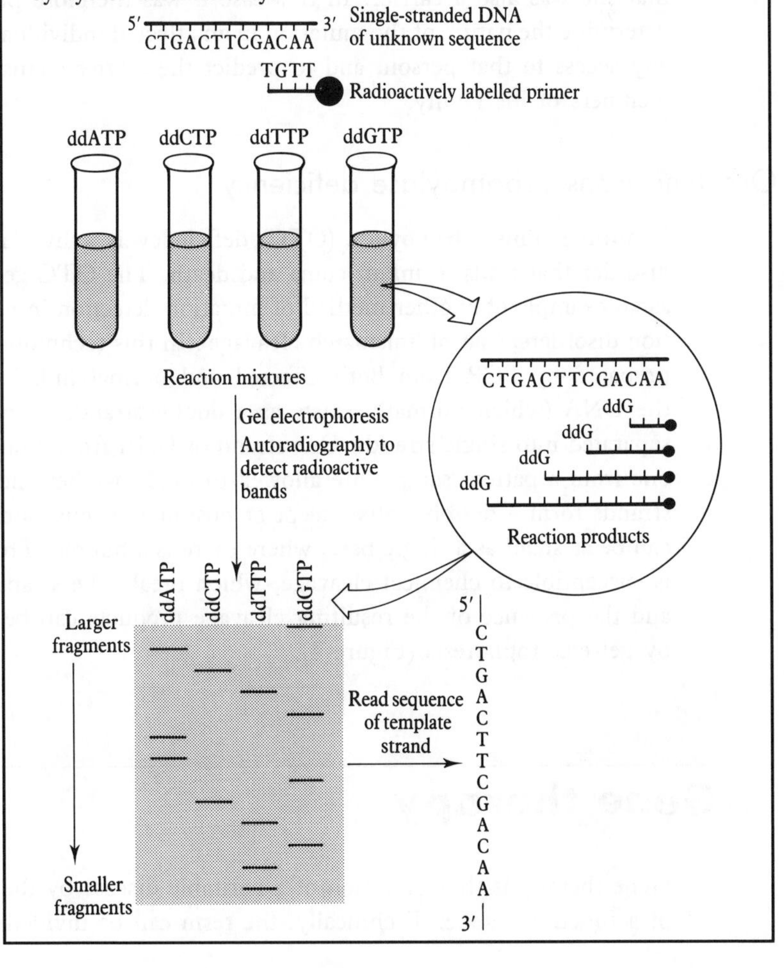

I will illustrate in detail one particular clinical example. A woman whose brother had died of Lesch–Nyhan syndrome wanted to know whether she was a carrier of the defective gene. No tissue sample of any kind from the deceased brother was available for analysis and it was therefore impossible to determine directly the mutation within his HPRT gene. The woman's mother was fortunately cooperative in the study and when her DNA was analysed by multiplex PCR and then by automated DNA sequencing, the sequence in one position was ambiguous, that is, two different bases were found at the same position. She was therefore assumed to be carrying one copy of a normal HPRT gene and one copy of the mutant gene responsible for her son's disease. It was then straightforward to search the DNA from the daughter for the same mutation which, since it was present, indicated that she was also a carrier. In this case it was therefore possible to determine the nature of the mutation in an affected individual without any access to that person, and to predict the carrier status of other members of his family.

Ornithine transcarbamoylase deficiency

Ornithine transcarbamoylase (OTC) deficiency is a liver metabolic disorder that leads to infant coma and death. The OTC gene serves as an example of another method of mutation detection in new mutation disorders, that of 'mismatch cleavage'. In this technique DNA is prepared by PCR from both affected and normal individuals and the DNA (which normally exists as a double-stranded molecule) is separated into single strands. One strand of DNA from a normal and one from a patient sample are allowed to bind together and the two strands form a double helix except at positions of mismatch, which can be as small as a single base, where there is a bubble. This bubble is susceptible to chemical cleavage, which breaks the strand in two, and the presence of the resulting cleavage products can be detected by gel electrophoresis (Figure 4).

□ Gene therapy

Gene therapy is the correction of a heritable disease by the addition of a functional gene. Technically, the term can be divided into two

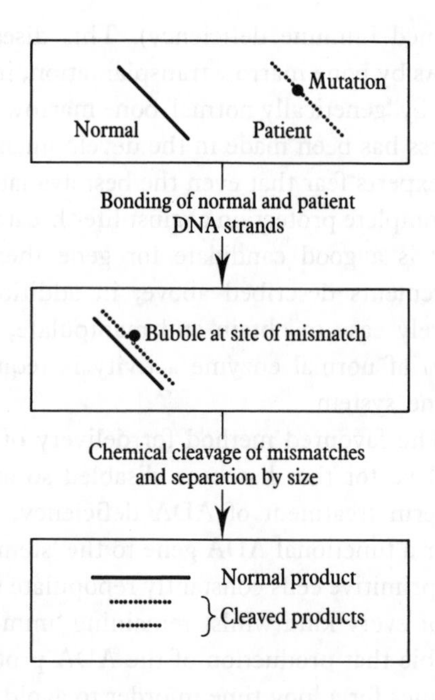

Figure 4 Diagram of chemical cleavage method of mutation detection. Single strands of DNA derived from the appropriate disease gene in normal subjects and patients are generated and bind together, forming hybrid double-stranded molecules. At points of mismatch (that is, sites of mutations) the DNA is cleaved chemically and the shorter products are detected by gel electrophoresis.

categories, namely 'germ-line gene therapy' and 'somatic cell gene therapy'. On ethical grounds, the latter is the only kind of gene therapy being considered in humans and involves the treatment of different cells in the body, but does not allow inheritance of the genetic changes; it is analogous to an organ transplant. For a disease to be a candidate for gene therapy the biochemistry of the disorder must be well understood, a cloned functional gene must be available and there must be means available to transfer that gene into a tissue where the gene will have some benefit. Current conventional treatment protocols must be nonexistent or inadequate.

Adenosine deaminase deficiency

Adenosine deaminase (ADA) deficiency is exhibited by about one-third of children completely lacking an immune system (SCID; severe combined immune deficiency). This disease can be cured in some patients by bone marrow transplantation, indicating that it can be corrected by 'genetically normal' bone marrow cells. Although remarkable progress has been made in the development of alternative treatments, most experts fear that even the best available treatments will not provide complete protection against life-threatening infections. ADA deficiency is a good candidate for gene therapy because it fulfils the requirements described above; in addition, bone marrow tissue is relatively easy to obtain and manipulate, and it is known that only 5–10% of normal enzyme activity is required for restoration of the immune system.

The favoured method for delivery of a functional ADA gene is a viral vector that has been disabled so as not to be infectious. For long-term treatment of ADA deficiency, it would be imperative to deliver a functional ADA gene to the 'stem cells' of the bone marrow; these primitive cells constantly repopulate the marrow with new blood cells of every kind whilst remaining 'immortal' themselves. It is also desirable that production of the ADA protein from the inserted gene continues for a long time in order to avoid the need for frequent bone marrow treatments.†

To summarize briefly the mouse experiments that have been performed in order to refine methods for gene therapy of ADA deficiency: a mouse can be lethally irradiated and then 'rescued' by supplying it with bone marrow infected with a virus vector carrying the human ADA gene. Mice rescued in this way still produce human ADA six months after the bone marrow transplant and marrow from the treated mouse can be used to rescue a second irradiated mouse, indicating that infection of stem cells has been achieved (otherwise repopulation of the second mouse's marrow would not be possible).

Human studies of ADA gene therapy have focused on treatment of human bone marrow kept in long-term culture and the results have so far been very encouraging. A functional ADA gene can be introduced into the most primitive cells detected in culture; these cells can give rise to others still detectable after 8–9 weeks in culture.

† *Editors' note* In 1992/1993, gene therapy using inactivated viruses to deliver 'useful genes' has been dramatically expanded to include the treatment (now being trialled) of some (skin) cancers, where the gene delivered to a tumour draws the attention of the body's immune system to the tumour and, therefore, to impede or perhaps to reverse its growth. This approach has the additional merit of not being dependent upon long-term repeated therapy, necessary for correcting heriditary deficiencies. Nevertheless, with respect to Dr Caskey's ADA example, gene therapy clinical trials are in progress in the USA with initially encouraging results, and limited trials for gene therapy for cystic fibrosis (CF) were approved at the end of 1992 by the US National Institute of Health.

Moreover, preliminary studies with normal and ADA-deficient bone marrow indicate that the transferred ADA gene is expressed at levels comparable to normal. The techniques of high-efficiency gene transfer into human cells and purification procedures for bone marrow stem cells are presently being combined with the aim eventually of administering to the patient his or her own, genetically-corrected, bone marrow stem cells. This would decrease the number of cells to be manipulated for a clinical application and would also decrease the risk of transplanting cells with 'undesirable' sites of integration of the virus vector. Although no instance of complication has been observed in mice following transplantation with bone marrow cells infected with a disabled virus, there is a concern that these viruses might in rare situations activate the development of cancer. Therefore the combination of gene transfer and stem cell purification procedures will result in safer protocols of gene transfer.

Ornithine transcarbamylase deficiency

There is a mouse model for the human OTC deficiency disorder, called the 'sparse fur mouse'. The defect in mice has been corrected by gene therapy, although the method used – germ-line gene therapy, for example, the insertion of the gene into the chromosome of an egg – is not an option in humans. Nevertheless, the vectors used for delivery of the functional gene in mice clearly work. Interestingly, expression of OTC in the small intestine was sufficient to correct a liver disorder. This enables us to consider the use of gene therapy in human liver genetic disease by treating the small intestine, a reasonably accessible tissue.

Duchenne muscular dystrophy

A recent development in our laboratory has been the assembly in the test tube of a single DNA molecule that is capable of expressing the large product of the DMD gene. This molecule is too big to be carried by the viral vectors currently used in other gene transfer experiments and so an alternative method of delivery must be found before gene therapy of DMD can be considered.

Ethical considerations

I will conclude this chapter with some illustrations of issues that need to be resolved in this era of advancing molecular medicine and the Human Genome Project. None is actually a new issue, but they will be raised more frequently in future. There are more questions than answers.

Diagnosis of carrier status

A family had one brother affected with DMD and two sisters with a good understanding of the disease and in close communication with their brother. They talked at length about his perspective on the disease and his attitude towards avoidance of the disease in his sisters' children. A second family had two sisters, one of whom bore a boy affected with DMD; the other had three daughters. The two sisters had a poor relationship early in life and that distance increased over time; the three daughters were never informed of the genetic risk of their transmitting the disease until they reached childbearing age. Both these families requested clinical study of the DMD carrier status of the females, but clearly the first family was in a much better position to understand the implications of the new information and to use it wisely. Preparedness and education in the general population are necessary for comprehension of genetic information.

In Italy, a screening program has been carried out since 1974 to detect carriers of β-thalassaemia, but unfortunately the information has not been kept confidential. Some of the young ladies in the small villages were known in the community to be carriers of the disorder and were therefore discriminated against in their selection for the formation of families. Again, education on the implications of being a carrier of a genetic disease is required, together with confidential treatment of the information.

In the Baltimore area, screening for carriers of Tay–Sachs disease amongst orthodox Jews is carried out (Jews have a higher incidence of this disorder than the general population). The resulting information is withheld from the individuals themselves, but made available to the matchmaker of the community, who then avoids the high-risk pairing of two carriers in the arrangement of marriages. This is another way of handling carrier testing information.

Major carrier testing for carriers of cystic fibrosis will shortly be initiated following the remarkable discovery of the major mutation causing the disease. However, this particular defect accounts for only 70–75% of all cystic fibrosis mutations and thus about one couple in 13 in the population will consist of one partner carrying the known mutation while the other has either a normal gene or an undetected mutation. Which of these two scenarios is the case will clearly affect calculation of the risk of producing a child suffering from cystic fibrosis. Testing for carrier status will always carry an element of the unknown; a negative testing for carrier status does not offer a guarantee that the individual does not possess a different mutation.

Adult onset disorders

The ability to diagnose a disease genetically before symptoms are apparent presents particular advantages and difficulties. One example of such a disease is Huntington's chorea, which becomes manifest in middle age and is transmitted by an affected person with a probability of 50%. Huntington's disease gene has not yet been cloned; but very accurate predictive tests have been developed.† Unfortunately, no therapy is available for Huntington's disease. So, is the information that a particular person will develop Huntington's disease a benefit (with respect to planning and reproductive decisions, for example) or a burden (in terms of worry, insurance coverage and so on)?

Type II hypercholesterolaemia individuals are at increased risk of cardiovascular problems, but if they could be identified before symptoms appeared, it would be possible to reduce the incidence of associated disease by drug treatment, diet and medical care. In this case it might be argued that there is clearly a benefit from possessing such knowledge, since this enables preventive measures to be taken.

Selective pregnancy termination

Should a couple be allowed the option of selecting the sex of their child by elective termination of a pregnancy of the 'wrong' sex? It is

† *Editors' note* In early 1993, the location of the Huntingdon disease gene now indicates the imminent cloning and complete characterization of the gene.

a very simple matter technically to determine the sex of a foetus: whether to apply the technology therefore becomes a policy decision. Legislation in the United States now prohibits the practice of sex selection. Let me describe another hypothetical situation. A couple are at risk of transmitting Lowe syndrome, which has an X-linked recessive inheritance; this means that a son would have a 50% chance of inheriting the disease, while a daughter would be unaffected (although she would have a 50% chance of being a carrier). Without more accurate genetic information about the status of the foetus the only means of avoiding transmission of the disease would be to terminate any male pregnancy, that is, sex selection, without knowing whether the foetus was affected or not.

Another issue concerns paternal identity, which can now be accurately determined using the 'DNA fingerprinting' technique described in Chapter 4. In two recent cases women who were trying to achieve a pregnancy under the care of an obstetrician were both criminally raped and subsequently became pregnant. Each has requested that the identity of the father of their foetus be determined. Is this application of recent technology reasonable or unreasonable? Consider the case of a woman who voluntarily consents to sexual relations with more than one man and asks the same question. If the technology is applied in one situation, should it be applied in the other?

□ Conclusions

In conclusion, recent advances in medical genetics have opened up new opportunities for the understanding and diagnosis of genetic disease. It is true that several issues concerning how best to disseminate and use genetic information have yet to be decided but that should not be an argument for blocking progress. Instead, we should anticipate the problems and be prepared to deal with them.

If we had rigidly followed the recommendations of the Asilomar meeting in 1975 that laid down rules guiding research in recombinant DNA experiments, we would never have allowed biotechnology companies to be established and products widely available on the market today would not be available. One example is insulin, the supply of which (from animal sources) for diabetics would be inadequate were it not for the manufactured recombinant product. Another example is that of growth hormone, which used to be prepared from cadavers but was then withdrawn because recipients began to develop viral central nervous system infections. The production of growth hormone from genetically engineered bacteria now ensures supply of a safe product. Let these serve as examples of the benefits of progress in genetics.

Acknowledgements

C.T.C. is a Howard Hughes Medical Institute Investigator. The assistance of Belinda Rossiter in the preparation of this manuscript is appreciated.

References

Chamberlain J.S. *et al.* (1990). In *PCR Protocols: A Guide to Methods and Applications* (Innis M. *et al.*, eds.) pp. 272–81. Orlando: Academic Press

Rossiter B.J.F. and Caskey C.T. (1991). Molecular studies of human genetic disease. *FASEB Journal*, **5**, pp 21–7

Watson J.D., Tooze J. and Kurtz D.T. (1983). *Recombinant DNA. A Short Course*. New York: W.H. Freeman

Weatherall D.J. (1991). *The New Genetics and Clinical Practice*, 3rd edn. Oxford: Oxford University Press

4

Genetic Fingerprinting: Applications and Implications

Professor Alec J. Jeffreys

Alec Jeffreys is a Professor of Genetics at the University of Leicester. He joined the department of genetics in 1977 with a main research interest in the molecular biology of haemoglobin genes. His early career was mostly focused on the evolution of genes in higher animals, but he is best known for his discovery of 'genetic fingerprinting'. This has revolutionized forensic science and has wide implications, not only for biology and medicine, but also for the legal profession.

With the exception of identical twins, we are each of us genetically unique. Modern molecular genetics provides the tools for exploring this uniqueness at the most fundamental level, namely variation in the genetic material, DNA. While most of our DNA varies little between people, there are ''stuttered'' regions of extreme variability that provide a method for uniquely identifying an individual by DNA fingerprinting. I will describe how DNA fingerprinting works and how it provides for the first time a method for positive identification in forensic and parentage analysis. In the past few years, DNA fingerprinting has been applied with extraordinary speed and great success to civil and criminal casework.

I will discuss the types of investigation now made possible by DNA typing, particularly using recent advances in molecular genetics, as well as the need for public reassurance about the scientific and operational probity of genetic fingerprinting. I will discuss the social and legal issues raised by DNA typing and will describe some of the other applications of genetic fingerprinting in, for example, animal breeding and the conservation of endangered species.

☐ Biological evidence in forensic science

Much of forensic analysis is concerned with the identification and matching of forensic evidence, whether in ballistics, hair and fibre analysis, drug characterization, paint identification, fingerprinting or suspect identification in identity parades. Frequently, the forensic scientist is asked to characterize biological remains from the scene of a crime and to match these specimens with potential suspects. Such evidence can include blood stains, semen recovered from rape victims, decomposed corpses, body parts, hair and saliva. Similarly, biological evidence of kinship may be required in civil cases (paternity disputes, immigration disputes) and criminal cases (rape or incest resulting in pregnancy).

Traditionally, such biological evidence has been subjected to genetic analysis using body substances, such as blood, serum proteins and enzymes, that show inherited variation between people. While such analyses are relatively simple and inexpensive, and can provide valuable evidence excluding a criminal suspect or establishing non-paternity, they suffer from several serious drawbacks. First, most of these genetic markers show only modest levels of individual variation and are therefore incapable of providing a unique biological identifier. For example, the genetically controlled ABO blood group system can be used to classify people into only four different types (blood groups A, B, AB and O), each of which is common in the UK. Matching of the ABO type between a forensic blood stain and a suspect therefore provides only weak statistical evidence for a true association. In parentage analysis the situation is worse. For example, if a mother belongs to group O and her child to group A, then the Mendelian laws of genetics dictate that the true father must have contributed to the child the gene for blood group A, and that he must therefore belong to group A or group AB; half of the UK population falls within these two groups and thus there is a 50% chance that a falsely accused man would fail to be excluded by this test. To circumvent these problems, forensic serologists have developed dozens of different markers which, together, can provide a good level of individual specificity. However, most of these markers are based on blood group substances that are not present in other body tissues and can therefore be used only to type blood. Furthermore, these markers are complex biochemical substances, which are unstable and frequently deteriorate in forensic specimens. As a result, biological evidence from the scene of a crime is often refractory to typing, or, worse, may even give spurious typing results that could lead to the false exclusion of a criminal suspect. In establishing paternity, sequential testing with many markers can exclude the majority of falsely accused non-fathers but cannot provide positive proof of paternity. (Indeed, several cases have emerged where the full battery of traditional tests has led to a conclusion of paternity, whereas subsequent DNA analysis has revealed a clear and unambiguous exclusion.)

These traditional genetic markers make use of variations in biochemical substances, such as proteins, that are controlled by inherited variability in the DNA that codes for them. Recent spectacular advances in molecular genetics have enabled us to explore DNA variation directly, which has led to the development of a range of enormously powerful DNA typing systems that have revolutionized not only forensic medicine but also many other fields of biology.

□ What is DNA?

The human body comprises trillions of cells, each of which (with the exception of red blood cells) contains a full set of chromosomes. There are 46 chromosomes per cell, 23 inherited from the mother and 23 from the father. Each chromosome is essentially a packet of DNA, an enormously long, thin molecule that carries the inherited information required for the development of an individual. The DNA molecule consists of the famous Watson–Crick double helix with two intertwined complementary or 'mirror image' strands (see Chapter 2, Box 1 p. 16. DNA can replicate by separation of these strands and synthesis of the missing complementary strand to produce two identical copies of the double helix, thereby ensuring that copies of the genetic material can be inherited from cell to cell and generation to generation. Each DNA strand consists of an almost endless chain of four different chemical building blocks, termed 'bases', genetic information being stored in the precise chemical sequence of bases along the DNA strands. Each cell contains a full set of 6 000 000 000† bases forming

† *Editors' note* The full set of genes (diploid complement) has 6 000 000 bases, whereas the half set that is passed on through sperm and eggs (the haploid complement) has 3 000 000 000 bases.

the 'book of life', the complete set of instructions for a human being. One great task of modern molecular genetics is embodied in the Human Genome Project, whose goal is to determine the complete chemical sequence of human DNA, an important step towards full understanding of the genetic control of development and disease in man. The Human Genome Project is discussed by Dr James Watson in Chapter 2.

There is no question that each of us, except identical twins, is a genetically unique individual and that this uniqueness stems from variation in the precise chemical sequence of bases along our DNA molecules. Since the late 1970s, molecular genetic techniques have allowed us not only to isolate and characterize human genes, but also to explore directly this most fundamental level of genetic variability, namely DNA variation. Some of this variation is injurious and is directly responsible for such inherited diseases as cystic fibrosis and Duchenne muscular dystrophy. Professor Tom Caskey in Chapter 3 describes how molecular genetics can reveal the basic DNA defects responsible for many inherited diseases and how this knowledge can be used to provide DNA-based diagnostic tests for such diseases. Such deleterious variation is, however, relatively rare in our DNA, and the great majority of DNA differences now documented appear to be 'silent' and without effect on the individual. This is not surprising since there is good evidence that most of our DNA appears, remarkably, to be a non-functional accumulation of evolutionary debris rather than coding DNA. This silent variation in our DNA has provided us with a novel and almost limitless source of new genetic markers, which are proving invaluable in the genetic mapping of genes along our chromosomes, including genes involved in inherited disease. However, the level of variation between individuals over most of our DNA is rather modest, and these DNA-based genetic markers suffer, as do most traditional markers, from a lack of informativeness.

□ What is DNA fingerprinting?

While most of our DNA shows little variation, regions called 'minisatellites' scattered along our chromosomes can show extreme levels of variability. These minisatellites consist of 'stuttered' regions of DNA, in which a short chemical sequence of bases is repeated over and over again. If, for example, a normal DNA sequence is written ... ABCDEFGH ... then a minisatellite would appear as ... ABCDEDEDEDEFGH ... with multiple copies of the motif DE, variability arising from differences in the number of DE repeats. As a result, some minisatellites can show dozens or even hundreds of different length states and thus provide the most variable and informative genetic markers yet discovered.

In 1984, we showed that the repeat units (DE in the above example) of different minisatellites tend to share a similar sequence motif, which seems to predispose DNA towards this stuttering. Discovery of this shared motif allowed us to design a method for highlighting many of these minisatellite regions simultaneously in human DNA, thus producing the first DNA or genetic 'fingerprint'.

DNA fingerprinting is a complex process, taking perhaps a week to proceed from biological sample to final result. The first stage is to extract chemically the DNA from suitable sources such as blood (DNA is present in white blood cells), semen, hair roots or mouth swabs; other sources of DNA, such as saliva and urine, generally yield too little DNA for fingerprinting by this method. Next, the extracted DNA is checked to ensure that sufficient good-quality DNA has been recovered for subsequent typing. It is then cut with a restriction enzyme, a protein that cleaves the DNA strands at specific positions, to produce a complex set of millions of different DNA fragments, some of which contain the variable minisatellites. The length of these minisatellite fragments is determined by the number of repeats or stutters, and the next stage is, therefore, to separate the DNA fragments according to length by passage through a slab of gel in an electric field. The pattern of DNA fragments sorted by size is then transferred from the gel to a sheet of membrane, which is subsequently treated to separate the two strands of the double helix within each DNA fragment without disrupting the pattern on the membrane. Next, the membrane is reacted with a radioactive 'probe', a segment of stuttered DNA which seeks out and forms a double helix with any minisatellite fragments on the membrane. As a result, the variable minisatellites become radioactive and can be visualized on X-ray film (see Chapter 3, Box 1 p. 33).

The end result of this process is a pattern of 30 or so bands or stripes on X-ray film, resembling to some extent the bar code used on supermarket goods. These film patterns, examples of which are shown in Figure 1, have three critical properties:

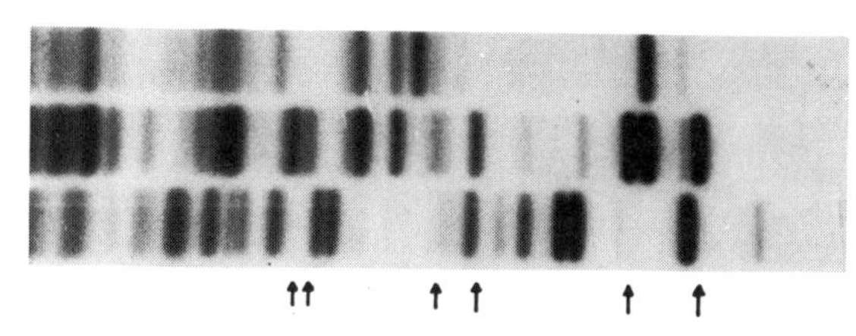

Figure 1 DNA fingerprint analysis of a paternity case. The child's pattern contains a number of bands not present in that of the mother, which must therefore have been inherited from the child's natural father. Most of these bands are not present in Mr Y (arrowed bands), proving that he could not be the child's father.

(1) The degree of pattern variation between individuals, even if they are closely related, is so extraordinary that we can legitimately refer to these patterns as DNA fingerprints, a unique biological identifier. Extensive comparisons of different people's patterns have shown that the odds against two people (other than identical twins) having the same pattern is remote in the extreme, to the extent that it would be very unlikely that *any* two people on earth would by chance share the same pattern.

(2) Despite this extraordinary variation, an individual's DNA fingerprint is essentially constant, irrespective of the source of DNA (blood, semen). Thus DNA fingerprinting can be applied to any appropriate source of DNA.

(3) DNA fingerprints show a simple pattern of inheritance, a child receiving approximately one-half of its bands from the mother and the remainder from the father. Occasionally, a new mutant band unattributable to either parent will appear in a child (these mutations are the ultimate source of this extraordinary variability), but extensive analysis of families has shown that the frequency of such mutant bands is low enough not to interfere significantly with the use of DNA fingerprinting in establishing family relationships.

Applications of DNA fingerprinting

Even though DNA fingerprinting was initially developed to provide a source of super-informative genetic markers for medical genetics, it was immediately apparent that this new technology could in principle be applied to forensic analysis to allow for the first time the positive linking of a forensic specimen and a criminal suspect. A report in 1985 on the successful isolation of (admittedly poor-quality) DNA from a 2400-year-old Egyptian mummy further suggested that DNA, unlike blood group markers, was remarkably tough and might survive in typable form in forensic specimens. Collaboration with forensic scientists at the Home Office Central Research Establishment, Aldermaston, quickly showed that biological stains, even when 5 years old, could sometimes be amenable to DNA fingerprinting. In particular, sperm DNA could frequently be typed from semen stains or vaginal swabs taken from rape victims, providing for the first time a method of unequivocally linking a rape victim with her assailant.

This form of DNA fingerprinting has now been successfully used in a number of criminal investigations, and since 1987 has been admitted in evidence in UK courts. It is remarkable that our law courts have accepted DNA evidence so rapidly, in view of the novelty of the test systems and the fact that the courts were being asked to consider evidence derived from a scientific field, molecular genetics, of which they had no experience. Fortunately, the scientific principles underlying DNA fingerprinting are not complex. Experience has shown that judges, lawyers and juries can readily understand the issues involved, a process which is greatly helped by the relatively simple and pictorial nature of the evidence.

Although DNA fingerprinting has been used successfully in a number of cases involving the analysis of forensic samples, technical considerations have led to its widespread replacement by single-locus probe analysis (see below), primarily because of the improved sensitivity of single-locus probes and the occasional difficulty experienced in comparing DNA fingerprint patterns obtained on different days and therefore on different sheets of X-ray film.

DNA fingerprinting has had a far greater impact on the determination of kinship, where obtaining good-quality blood DNA is not a problem and where all relevant evidence (DNA fingerprints from the alleged mother, child and father) are produced side-by-side on the same piece of X-ray film, eliminating the problems associated with inter-run comparisons. One major application of kinship testing is the resolving of paternity disputes, normally in civil cases, but also occasionally in criminal cases, such as rape or incest, which result in the victim becoming pregnant. UK civil and criminal courts have fully accepted the validity of DNA evidence in such cases, and indeed a substantial proportion of UK paternity disputes are now resolved by DNA fingerprinting. Given the unequivocal nature of the evidence, many paternity disputes are now being settled out of court, with consequent significant savings in time and expense both to the disputants and to the courts.

The second major application of parentage analysis is in the resolution of immigration disputes that involve uncertainties about claimed family relationships between the UK sponsor and his alleged wife and children abroad; such cases primarily originate from the Indian subcontinent. Indeed, the first case in which DNA fingerprinting was applied, in April 1985, was an unusually difficult immigration dispute involving a boy who had left his family in the UK and had subsequently returned from abroad. Irregularities in his passport suggested to the immigration authorities that a substitution had occurred, and that the returning boy was either not related to the family or was a cousin of the other children in the family. Conventional blood group testing supported the boy's claim, but not conclusively. We were therefore asked by the family's lawyer to attempt a DNA analysis of the family. Unfortunately, the father was not available, but we could circumvent this problem by comparing the DNA fingerprints of the mother and her three undisputed children to enable us to reconstruct most of the missing father's DNA pattern. Comparison of the disputed boy's DNA fingerprint with those of his alleged mother and (missing) father showed beyond any doubt that the boy was an authentic full member of the family. As a result of the DNA evidence, the case against the boy was dropped by the Home Office and he was allowed to return permanently to his family. Subsequent experience with immigration disputes has shown that the statistical power of DNA fingerprinting is sufficiently great to be able to resolve difficult cases in which, for example, a UK sponsor is suspected of attempting to bring to this country not his own wife and children, but those of his (unavailable) brother.

In view of the demand for DNA fingerprinting, which became more than evident during 1985 and 1986 when my laboratory carried out the first analyses in paternity and immigration disputes, ICI established in 1987 a subsidiary, Cellmark Diagnostics, to provide a DNA fingerprint service under an exclusive licence from the Lister Institute of Preventive Medicine, which holds the patent rights to DNA fingerprinting. Since its establishment Cellmark has processed many thousands of paternity, immigration and forensic cases. Its work and the DNA fingerprinting system have been recognized by the award of a British Standard, by the accreditation of Cellmark by the Home Office as blood testers, and by the recent Queen's Award for Industry gained jointly by Cellmark and the Lister Institute. In addition, powers have recently been granted to the courts enabling them to require the taking of blood for DNA fingerprinting in paternity disputes. Finally, the Home Office and the Foreign and Commonwealth Office have published a joint report on a pilot immigration study on Pakistani and Bangladeshi volunteer families carried out by the Home Office, Foreign and Commonwealth Office, Cellmark and myself which established the feasibility of routine DNA testing in immigration disputes.

According to the Home Office, in 1989 some 2000 immigrants were reunited with their families, thanks primarily to DNA evidence, including a significant number who had been previously refused entry as a result of Entry Clearance Officers' investigations. It is now clear that the great majority of immigrant families who voluntarily avail themselves of DNA typing have the relationship claimed. However, in the absence of available data on the proportion of applicants who refuse testing (and who would therefore be more likely to include individuals knowingly making false claims), it is not possible (for me at least) to estimate the overall proportion of applicants for immigration making false claims. *Prima facie*, one would have to conclude that this proportion is small, and that significant injustice has been done to many families previously refused admission by Entry Clearance Officers.

DNA fingerprinting in immigration disputes raises other issues. First, who should pay for the tests? My own view is that, since most family claims are authentic, the burden of proof ('innocent till proven guilty') and financial liability should rest with the immigration authorities, and ultimately the government, not with the frequently impoverished families. However, the centrally administered DNA scheme which came into force in 1991, is covered by the fees that all would-be immigrants must pay, irrespective of where they come from in the world and whether or not they require a DNA test.

Second, advice about DNA testing should be freely and fully available from the immigration authorities, through organizations such as the UK Immigrant Advisory Service and the Joint Council for the Welfare of Immigrants and through local community outlets such as Law Centres. Third, it is important that families are properly advised prior to testing, particularly since the point at issue is a rigorous biological definition of family relationships rather than the conventional criteria of marriage, cohabitation and child rearing. In particular, both the family and the immigration authorities should be aware of the implications of non-paternity associated with marital infidelity, where the husband genuinely believes himself to be the father of the child; under such circumstances the family is placed in double jeopardy, with the child failing to provide the genetic link between the claimed husband and wife, leading to refusal of entry, and the family potentially destroyed through the evidence of infidelity.

Single-locus probes and DNA profiling

The DNA fingerprinting system described above uses 'multilocus' probes, so called because each probe detects multiple minisatellites at different locations in human DNA to produce complex banding patterns. Using molecular genetic techniques, we and others have isolated individual minisatellites from human DNA to provide a battery of 'single-locus probes', each of which detects a single highly variable minisatellite region in human DNA. The resulting pattern produced by each probe on X-ray film is a much simpler 'DNA profile' consisting of two bands or stripes per individual, one band corresponding to the particular version of that minisatellite inherited from the mother and the other to that inherited from the father (Figure 2).

These single-locus probes are considerably more sensitive than their multilocus counterparts. In addition, they can provide information from partially degraded DNA refractory to multilocus analysis, and can detect mixed DNA samples (for example, semen from a victim of gang rape). The two-band patterns show good variability between unrelated individuals, but much less variability between close relatives, such as brothers and sisters. The pattern per probe test is

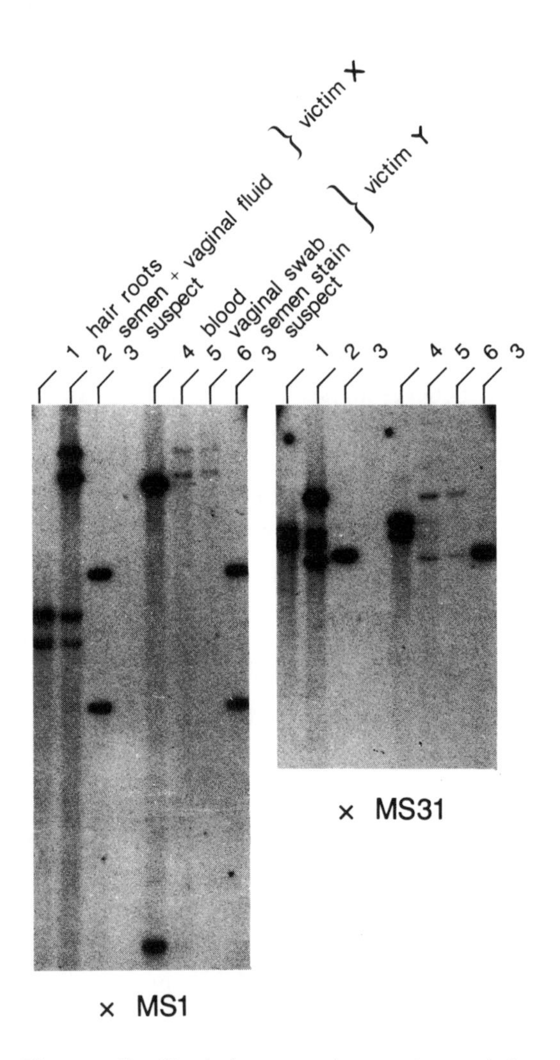

Figure 2 Single-locus probe analysis of forensic samples from the Enderby murder case. This case involved two girls who had been raped and murdered. DNA profiles were obtained from forensic material and from blood taken from a prime suspect, using two independent single-locus probes (MS1 and MS31). Each probe produces a DNA profile of two bands per individual or, occasionally, one band if the two bands are superimposed. Two semen DNA bands not attributable to the victim can be seen in samples 2, 5 and 6. The bands in samples 5 and 6 are faint since very little DNA was recovered from these specimens. The DNA profiles of semen recovered from victims X and Y are indistinguishable, providing strong evidence that both girls had been raped by the same man. The semen profile is different from that of the suspect, which immediately eliminates him as the source of the semen found on each victim.

not individual-specific (hence the term 'DNA profile', rather than DNA fingerprint), but good levels of specificity can be achieved by sequential testing with a panel of (typically) five different single-locus probes.

Single-locus probe analysis provides a powerful system for forensic testing, and was used in the first murder investigation to employ DNA analysis. This case (the Enderby murder case) involved two schoolgirls, both of whom had been raped and strangled. A young man was subsequently arrested, who confessed to the second murder. By DNA profiling of semen recovered from the victims, we showed that both girls had been raped (and therefore presumably murdered) by the same man, who was *not* the prime suspect (Figure 2). As a result, and following confirmatory analysis by Home Office forensic scientists, the young man was released in the first case of a man being proved innocent by DNA testing. Faced with the lack of any prime suspect, and having as evidence only the anonymous DNA profile of the assailant, the police elected to carry out the first mass screening within a local community using a combination of traditional genetic markers and DNA typing, a strategy which successfully flushed out the true murderer and heralded the beginning of forensic DNA analysis in the UK. It is worth noting that the mass screening strategy worked in this case only because there was a well-defined local rural community with a strong motivation to volunteer for blood sampling (either through a desire to see the murderer brought to justice or, in the case of waverers, through substantial social pressure within the community). For these reasons, it is most unlikely that mass-screening strategies would work in most cases, such as inner-city crime.

DNA profiling using single-locus probes is now established at the Home Office Central Research Establishment, regional Home Office forensic laboratories and both the Metropolitan and the Strathclyde Police Laboratories. A commercial UK service is provided by Cellmark Diagnostics and, using alternative probes, by University Diagnostics. In the United States, the FBI has been engaged in evaluating a panel of single-locus probes and is extensively applying the system to criminal investigations. State laboratories are adopting the tests, which are also provided by a number of US commercial laboratories, including Cellmark Diagnostics. At present, almost every major country in the world has either implemented DNA profiling or is preparing to implement the method.

Is DNA profiling reliable?

DNA analysis provides unusually powerful and persuasive evidence for courts to evaluate, since this evidence is frequently accompanied by biostatistical evaluations indicating that the matching pattern between the forensic specimen and defendant is exceedingly rare in the population, occurring perhaps in one person in a million, a billion or even a trillion. Given such odds, in conjunction with primary evidence that is simple, pictorial and can be readily appreciated by a jury (for example, in the form of X-ray film), there is a real risk that the expert witness could (albeit unintentionally) relinquish his or her role as a provider of information to be evaluated, together with other evidence, by the jury. Instead the jury could be asked to provide what would be tantamount to a definitive statement of guilt or innocence. Not surprisingly, in view of this, there has been considerable and legitimate discussion of the validity and reliability of DNA evidence, initiated in particular by the Castro murder case. In this US case, the commercial company Lifecodes provided testimony claiming the clear matching of DNA from a murder victim with blood found on the watch of the suspect. In a pre-trial hearing the defence discovered many serious shortcomings in the evidence, including poor-quality and overinterpreted DNA profiles and inconsistencies in the biostatistical evaluations of the evidence. As a result, the evidence was dismissed, although the defendant subsequently pleaded guilty to the murder. Some of the issues to emerge in this and subsequent discussions include:

(1) The public and the judiciary must have full confidence that testing laboratories are operating at the highest level of technical excellence and using rigorously tested standard operating procedures designed to minimize any possibility of sample mix-ups, with full quality control procedures in place. Laboratories should be open to external assessment, ideally by 'blind' trials (submission of dummy casework), such as those performed by the Metropolitan Police and the Home Office Laboratories and shortly to be offered on an international basis by Cellmark Diagnostics.

(2) Poor-quality DNA evidence should be replaced after retesting to give high-quality data or, if necessary, its shortcomings should be

fully and openly described in testimony. Note that technical imperfections will tend to generate apparent mismatches between the forensic evidence and the defendant (for example, impurities in the forensic DNA might shift the forensic DNA pattern out of alignment with that of the suspect). As a result, technical imperfections give results that are strongly biased in favour of the defendant, turning matches into apparent mismatches but not vice versa.

(3) No matter how careful the testing laboratory, the possibility of sample mix-ups, either in the laboratory or at the scene of the crime, always remains. However, a mix-up is far more likely to generate spurious mismatches, again exonerating the guilty rather than incriminating the innocent.

(4) DNA evidence must be interpreted in the context of other evidence, and not alone. For example, proof that a semen stain on the clothing of a rape victim matches the suspect does not by itself prove that the suspect was the rapist. First, had a rape actually taken place? If so, was the semen stain produced at the time of the rape? Second, is there any other scenario that could produce an innocent explanation of the semen stain?

(5) Could biostatistical evaluation of the weight of the DNA evidence be seriously distorted by genetic differences between ethnic groups or by inbreeding within local communities? Extensive data from my own laboratory, from the Home Office Central Research Establishment and from Cellmark indicates that such distortions are most unlikely to be a significant problem, at least within UK communities. In addition, all statistical evaluations are deliberately conservative to prevent excessive bias against the interests of the defendant in the weight of the evidence.

In the light of the Castro case and the subsequent controversy, the US Congress Office of Technology Assessment commissioned an independent report on DNA testing in the US, which concluded that the scientific validity of DNA analysis is beyond dispute but highlighted the need for standards in the testing laboratories. It is worth noting that probably never before has a forensic test come under such intense scrutiny, and it is a testimony to the power of the technology that it has withstood such a concerted attack, and indeed has emerged as a superior service after the elimination of several poor practices.

☐ Super-sensitive DNA typing

DNA profiling using single-locus probes requires good-quality DNA from at least 5000 cells, roughly equivalent to the amount of DNA recoverable from a single hair root. Frequently, degraded or minute forensic specimens yield insufficient DNA for typing. To circumvent this problem, Cetus Corporation in California has developed a method for copying and thereby amplifying DNA in the test tube, using a technique termed the polymerase chain reaction (PCR; see Chapter 3, Box 2 p. 35). This method can be used to amplify specific segments of DNA that are of interest to produce enough DNA for analysis. We and others have developed a range of amplifiable genetic markers, including minisatellites, much shorter 'microsatellites' and regions of true genes, such as those controlling transplant rejection (HLA genes), which are known to vary between individuals. With PCR, it is now possible under laboratory conditions to extend DNA typing to the quantum level of a single cell.

Given the extraordinary sensitivity of PCR DNA typing, the range of forensic applications could be substantially broadened to include frequently intractable evidence such as hair roots, urine and saliva; in the latter case, there is the real possibility, following further research, of DNA typing of saliva traces on documentary evidence such as blackmail letters and letter bombs. PCR DNA typing can also be used to analyse severely degraded DNA in, for example, decomposed forensic samples. Indeed, recently we successfully identified the skeleton of a murder victim buried for eight years by comparison of DNA extracted from bone with DNA from the parents of the presumed victim. Retrospective analysis of criminal cases will also become possible, creating the real likelihood of a flood of appeals against criminal convictions once this technology has become generally available. However, PCR DNA typing should be treated with some caution in view of its extreme sensitivity, which makes it especially vulnerable to contamination, either at the scene of the crime or in the testing laboratory. This issue is not trivial, but one which can be addressed by appropriate technology.

Other applications of DNA typing

The principal applications of DNA typing, which have attracted the most public attention, are in forensic analysis, paternity disputes and immigration cases. However, DNA marker systems have found many other actual or potential uses. These include:

- The development of DNA markers for studying the genetic basis of inherited disease and cancer. Single-locus minisatellite and microsatellite markers provide some of the most informative genetic markers in man and are proving invaluable in constructing genetic maps of human chromosomes. These maps make it possible, by studying families affected by an inherited disease, to localize the defective gene to a specific chromosomal region, which is the first stage in isolating the disease gene itself. Similarly, cancer arises when a DNA change or a gross chromosomal defect disrupts the normal growth properties of a cell, causing uncontrolled cell growth and the formation of a tumour. DNA markers have been invaluable in studying the DNA changes occurring in tumours, and have helped in the identification of some genes which, when mutated, are directly involved in tumour development.
- The identification of identical twins. Identical or monozygous twins derive from the same fertilized egg and therefore share the same DNA. In contrast, non-identical or dizygous twins develop from separate fertilized eggs and are no more like each other genetically than are other brothers and sisters. DNA fingerprinting easily distinguishes monozygous from dizygous twins. The results of such tests are frequently of great interest to the twins themselves. Twin studies and therefore the reliable identification of identical twins are also important in studying the genetic basis of complex human characteristics, such as height, weight and predisposition to disease, since any difference between identical twins, who share the same DNA, must have an environmental, not a genetic, basis. Twin testing is also important in transplantation surgery, since an identical twin provides an ideal tissue donor in that transplant rejection due to tissue mismatch is prevented.

- Monitoring the success of transplants. The main clinical application of DNA typing to date is in bone marrow transplants, most frequently in people being treated for leukaemia. In this operation, the patient's own cancerous bone marrow is destroyed by a combination of radiation and chemical therapy, and replaced by normal bone marrow from a healthy tissue-matched donor. Since blood cells are made in the bone marrow, the success of the operation can easily be monitored by blood DNA typing to check that the DNA in the circulating blood is that of the donor, not the patient. Any reappearance of the patient's own DNA in the blood might signal the reappearance of cancerous cells, and appropriate therapy can then be instituted.
- Animal breeding. Remarkably, the multilocus probe systems developed for human DNA typing also produce highly variable and informative patterns from a wide range of animals, birds, reptiles, amphibians and fish, and in some cases even from invertebrates. Single-locus minisatellite and microsatellite markers are also being developed from non-human species. DNA typing is already providing animal breeders with a powerful new tool. For example, stolen animals can be identified from their DNA fingerprints. Similarly, DNA typing can be used to verify the identity of semen used in artificial insemination programmes, and also to establish the pedigree of an animal. Indeed, several cases involving disputes over whether a champion stud dog had really sired a given puppy have been satisfactorily resolved by DNA fingerprinting. In the long term, DNA markers will make it possible to construct genetic maps of domesticated animals and thereby enable the eventual localization of genes controlling economically important traits, such as disease resistance, milk yield and body weight. These markers will have a dramatic effect on the ability of animal breeders to improve animal stocks by artificial selection.
- Conservation biology. DNA typing is already helping to protect endangered species in various ways. It provides for the first time a method of identifying animals and birds stolen from the wild. This can be achieved either by setting up a database of DNA fingerprints of wild animals or birds against which a captive individual can be compared, or by showing that young individuals held by a breeder could not be the offspring of any other individuals that the breeder has in stock. A second application is found in helping zoos in their breeding programmes, in particular by identifying closely related individuals and thereby minimizing

the risk of inbreeding. More generally, DNA typing is beginning to revolutionize our understanding of the genetic makeup and breeding systems of natural populations, knowledge which is of fundamental importance in monitoring the genetic diversity and reproductive success of natural populations of animals and birds.

- Plant breeding. Although in its early days, it is clear that DNA typing can be extended to plants, and should be able to provide the plant breeder with useful breeding tools, as discussed above. It might also be able to provide unique identifiers for commercially important strains or cultivars, thereby helping to identify and reduce the risk of strain thefts.

Further considerations in forensic analysis

Over the past few years molecular genetics has provided an extraordinarily powerful set of tools for the forensic scientist. The technology is still rapidly evolving, and it is in the nature of scientific research that the range and application of tests five or ten years from now are largely unpredictable. For this reason it will be difficult for forensic laboratories to standardize tests completely, although interim standardization of a set of single-locus probes has already been largely achieved. Standardization is particularly relevant in the creation of DNA profile databases from population surveys, which provide the frequencies of patterns required for biostatistical evaluation, and for investigative purposes. The latter could include the DNA profiles of previous offenders (the significant rate of recidivism amongst offenders would undoubtedly make such a database a powerful tool in solving some cases of violent crime), unidentified bodies and relatives of missing persons. Ultimately, such a database could include the DNA profiles of everyone in the UK; while this would dramatically increase the ease with which some violent crimes could be solved, especially in cases in which there were no suspects, major economic, social and political issues associated with a global database make it most unlikely, in my view, that such a mandatory database would ever be constructed. In the light of this restriction, criminal investigations where DNA evidence is available will frequently flounder due to the lack of a suspect. It is important to note that a

DNA profile is fundamentally anonymous, and that no information concerning sex, ethnicity, disease status or physical appearance can be discerned from these patterns. Yet, ironically, all these features must be reflected in our DNA make-up, and the question arises as to whether DNA analysis could ever yield information contributing to a 'photofit' of the assailant. The sex of an individual can already easily be determined by other methods of DNA analysis. In the not too distant future, it is conceivable that DNA tests yielding information on, for example, ethnicity, hair colour and eye colour might become available. However, the idea that we could obtain a more complete physical description by DNA analysis is fanciful in the extreme, since our ignorance of how genes control physical appearance and behaviour is absolute, and could only be changed by a fundamental revolution in our ability to dissect such complex genetic characteristics in man.

Acknowledgements

I would like to thank my many colleagues who have helped in the development of DNA fingerprinting, and the Medical Research Council, the Lister Institute of Preventive Medicine and the Wolfson Foundation for supporting our research. DNA fingerprinting and some of its applications were originally described by the author, Wilson V. and Thein S.L. (1985), Hypervariable 'minisatellite' regions in human DNA, *Nature*, **314**, 67–74 and Individual-specific 'fingerprints' of human DNA, *Nature*, **316**, 76–9. More general discussions of the technology, its applications and limitations can be found in Kirby L.T. (1990), *DNA Fingerprinting: An Introduction*, New York: Stockton Press; Ballantyne J., Sensabaugh G. and Witkowski J., eds. (1989), *DNA Technology and Forensic Sciences*, Cold Spring Harbor, NY: Cold Spring Harbor Laboratory Press; and US Congress Office of Technology Assessment (1990), *Genetic Witness: Forensic Uses of DNA Tests*, Washington, DC: US Government Printing Office.

Further general details of DNA fingerprinting can be obtained from Cellmark Diagnostics, 8 Blacklands Way, Abingdon Business Park, Abingdon, Oxfordshire, OX14 1DY.

5

Prospects for Pre-implantation Diagnosis of Genetic Disease

Dr Anne McLaren

Anne McLaren was until recently Director of the Medical Research Council's Mammalian Development Unit in London. She is a leading figure in the study of mammalian genetics and embryology and is now at the Wellcome/CRC Institute in Cambridge. Pioneering molecular methods were carried out in the MRC Unit for the screening of genetic defects in embryos before they are implanted in the mother's womb. With the technique of *in vitro* fertilization, individual cells from the early embryo can be examined for the presence of a mutant gene. If an embryo at risk of developing a serious hereditary disease is found by this means to be free of the defective gene, it can be implanted. This type of selective implantation of healthy embryos reduces the necessity of introducing a normal gene into the sick embryo, that is, 'gene therapy'.

In principle, information about the genetic constitution of any oocyte can be obtained by analysis of the first or second polar body, while analysis of one or more cells removed during cleavage or at the blastocyst stage can yield information about a pre-implantation conceptus. If blastocysts could be removed by uterine lavage, *in vitro* fertilization would not be required. For couples at high risk of producing a child with a serious genetic or chromosomal defect, pre-implantation diagnosis offers advantages over prenatal diagnosis later in pregnancy.

Diagnoses can in principle be made either at the protein level, using enzyme micro-assays, or at the DNA level, using the polymerase chain reaction or DNA/DNA hybridization. Biopsy procedures are already being tested to ensure that they do not prejudice future embryonic development. Distinguishing male from female among the pre-embryos of women carrying harmful sex-linked genes, such as haemophilia or Duchenne muscular dystrophy, has recently been the earliest clinical application of pre-implantation diagnosis (see Editors' note p. 80). Preliminary attempts are also being made to develop methods of diagnosing β-thalassaemia, cystic fibrosis and alpha-1 anti-trypsin deficiency before implantation.

From the point of view of the 'high risk' couple, pre-implantation diagnosis would be preferable to diagnosis later in pregnancy, since it would avoid the necessity of deciding whether or not to terminate an affected pregnancy. From a broader point of view, pre-implantation diagnosis removes the temptation to experiment with human germ-line gene therapy, that is, the introduction of genes or the replacement of genes in eggs or sperm that will then be inherited by future generations.

Prenatal diagnosis

Embryonic development is an immensely complicated process. It is perhaps not surprising, therefore, that something like 10% of all babies born suffer from some defect, either caused by a genetic or chromosomal abnormality, or a congenital malformation arising during pregnancy (Table 1). Some of these defects are serious, while others are quite mild. We are concerned here only with genetic diseases due to single gene defects, which affect about 1% of liveborn babies. Many genetic diseases are very serious indeed: for example Tay–Sachs, where babies normally die before the age of two, Lesch–Nyhan, victims of which die in their early teens, Duchenne muscular dystrophy, haemophilia and cystic fibrosis. All these greatly reduce the expectation of life and also, of course, the quality of life.

For couples who are 'at risk', that is, where both partners (or for some diseases, just one partner) are carriers of the same defective gene, the chance of producing an affected baby is high: on average, one in two or one in four of their children will be affected. A couple may discover that they are specifically at risk either because they themselves have been diagnosed as carriers, or because they have already produced an affected child. In their next pregnancy, they may well decide to have a prenatal diagnosis. For genetic diseases, that usually means amniocentesis or foetal blood sampling in mid-pregnancy or, nowadays, chorionic villus sampling earlier in

Table 1 Incidence of hereditary or congenital defects among liveborn infants per 1000 live births.

Genetic defects:	
Dominant and X-linked	10.0[a]
Recessive	1.1[a]
Multifactorial and irregularly inherited	47.3[a]
Chromosomal abnormalities	6.2[b]
Congenital malformations	42.8[a]
Total abnormalities	107.4

[a] United Nations Scientific Committee on Effects of Atomic Radiations (1977). New York: United Nations

[b] Hook E.B. and Hamerton J.L. (1977). In *Population Cytogenetics* (Hook E.B. and Porter J.H., eds.). New York: Academic Press

pregnancy. Before these techniques were introduced, a couple who knew they were at risk of producing a baby with a severe genetic disease would often decide not to have a family at all, or might produce one affected child who died after a few traumatic months or years. They might well then decide to have no more children. With prenatal diagnosis, if the foetus proves to be normal – as three out of four are in the case of a recessive disorder, for example – the couple can stop worrying. But if the foetus is affected, in some countries the couple would have the option of terminating the pregnancy.

For some couples, abortion is nevertheless wholly unacceptable. For all couples, terminating a pregnancy is very distressing, and obviously more so when it is a very much wanted pregnancy. Even with early chorionic villus sampling, one still has to go through the first three months or so of a pregnancy to no avail, and those are usually the most disagreeable months. And, of course, some couples are unlucky. For example, a severe and painful blood disease, β-thalassaemia, has a particularly high incidence in people from Cyprus or other parts of the eastern Mediterranean, or from the Indian subcontinent. Couples in which both partners have been diagnosed as carriers know that they have a one in four chance of producing an affected baby. Dr B. Modell in London has cited two such couples in particular, who had five prenatal diagnoses and terminated four pregnancies at 20 weeks because the foetus was affected (Modell, 1986). That is surely no way to have a family. Other couples may decide to call a halt after two or three terminations. Unless the situation changes they will not try again. But the situation could change if pre-implantation diagnosis of genetic disease, avoiding altogether the trauma of terminating a pregnancy, became a clinical reality. Dr Modell asked a group of Greek Cypriot couples at risk of having a baby suffering from β-thalassaemia what would most improve their family situation. They replied: 'To be able to start a pregnancy feeling committed to it, in the knowledge that it will not be affected'. It is precisely this knowledge that pre-implantation diagnosis aims to give to 'at-risk' couples.

I shall now describe a little of the background to early human development and the pioneering achievements of *in vitro* fertilization, which form the basis for the use of pre-implantation diagnosis.

Early human development

What exactly is pre-implantation diagnosis? To grasp what is involved, it is necessary to know something of human embryonic development. Thanks to the pioneering efforts of Patrick Steptoe and Robert Edwards, *in vitro* fertilization (IVF) was introduced in the UK some years ago as a treatment for the alleviation of infertility; thanks to IVF, we now know a good deal about the early stages of human development.

In a normal woman, sperm pass up the fallopian tube from the uterus to the ovary and fertilize the eggs that come from the ovary; the fertilized eggs pass down the same tube to the uterus. But if the tube is blocked, sperm cannot reach the ovary, fertilized eggs cannot reach the uterus and the woman is infertile. In IVF the woman's ovaries are usually stimulated with hormones to induce more than just one egg to be matured, the eggs are recovered and fertilized outside her body with her partner's sperm, the fertilized eggs are incubated

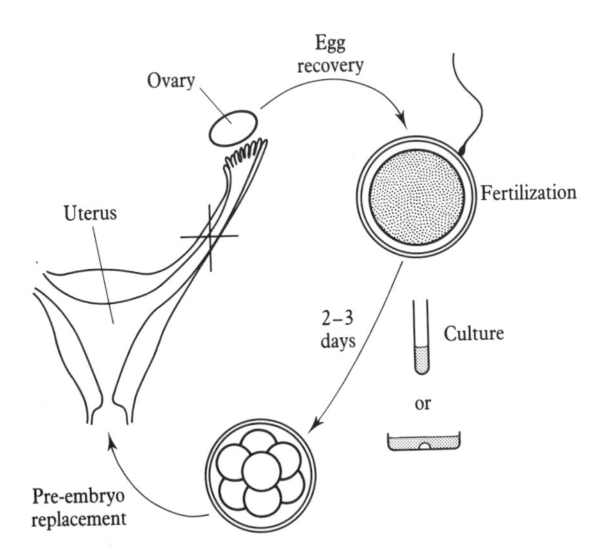

Figure 1 When the tube leading from the ovary to the uterus is blocked, so that the sperm cannot reach the egg and the egg cannot reach the uterus, pregnancy can still be achieved by recovering the egg from the ovary, fertilizing it outside the body, culturing it for a few days *in vitro* and then replacing it in the uterus. This procedure is known as IVF (*in vitro* fertilization). (*Source*: McLaren A., 1989.)

for two or three days in a glass or plastic dish to make sure that they are growing normally and the eggs are then replaced in the woman's uterus (Figure 1). After fertilization the egg divides into two cells, then four, eight and so on. This is the stage, as I shall describe, which is critical to the development of pre-implantation diagnosis of genetic defects in the fertilized eggs.

At the end of five days the so-called blastocyst stage has been reached, a clump of about 100 cells hollowed out with fluid in the middle. The blastocyst then begins to implant, that is, to burrow into the wall of the uterus. This process takes about a week. Implantation

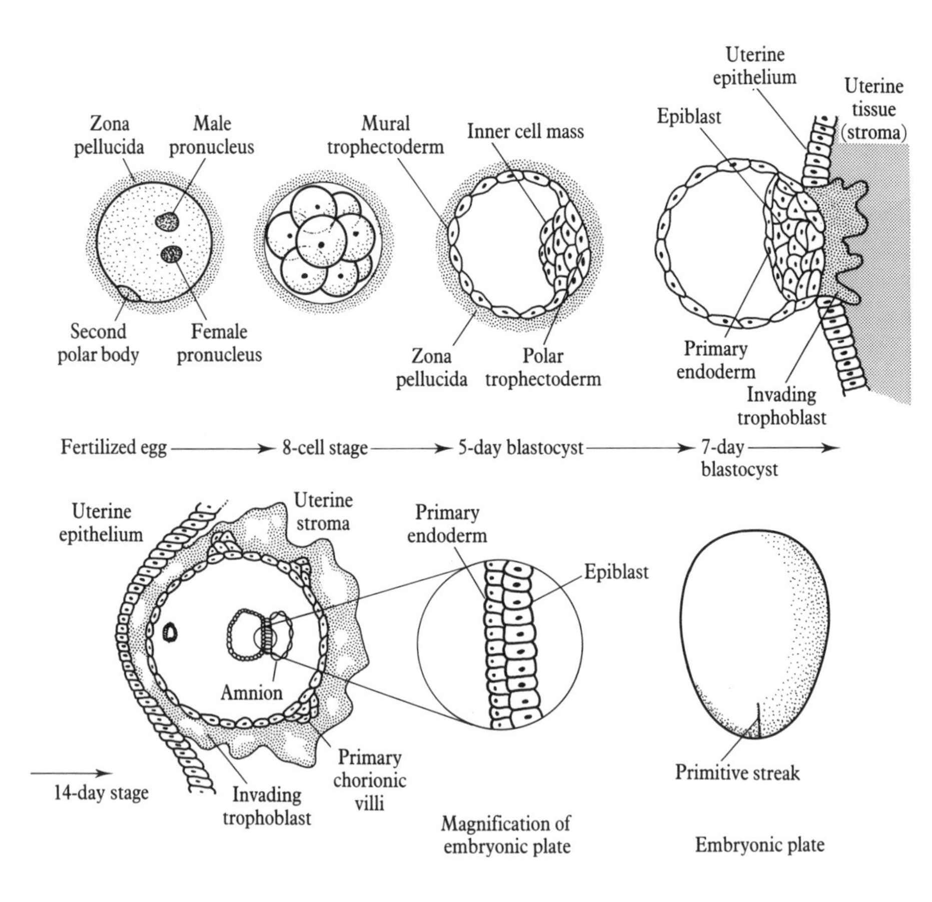

Figure 2 After about five days of development, the pre-embryo is a hollow ball of cells (shown here in cross-section). The actual embryo does not begin to form until the primitive streak disappears at about 14 days. (*Source*: McLaren A., 1987.)

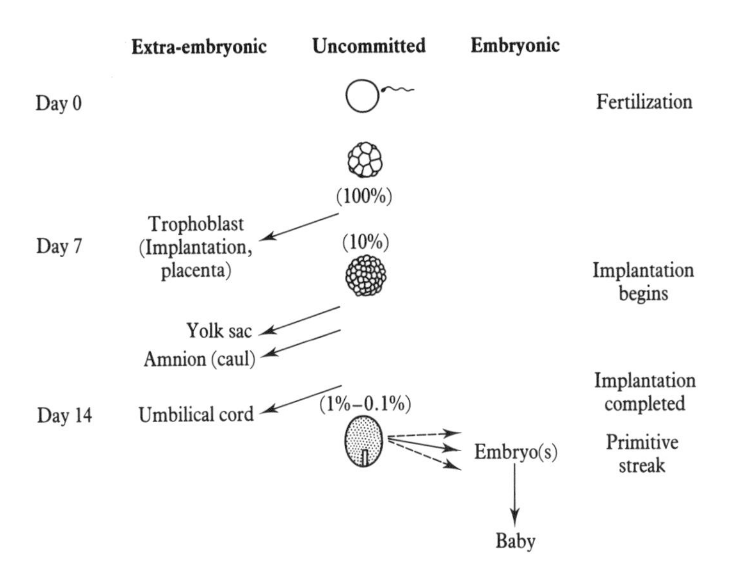

Figure 3 The first 14 days of human development. (*Source*: McLaren A., 1990.)

is completed about two weeks after fertilization (Figures 2 and 3). At this time 99% or more of the tissue derived from the fertilized egg is specialized to give rise to the life-support systems that are going to nourish and protect the future embryo. Only the very small number of cells that have remained unspecialized go on to produce the actual embryo itself. This is the so-called 'primitive streak' stage: it is around the 'streak' (or groove) that the embryo actually forms. Once the embryo has started to take shape, development proceeds very rapidly indeed (Figure 4). By eight weeks after fertilization, the foetus is fully formed, even though it is still only one or two centimetres in length.

□ Pre-implantation diagnosis of a genetic defect – what are the prospects?

In terms of pre-implantation diagnosis, only the first few days of development are relevant: that is, before implantation has begun. The strategy for pre-implantation diagnosis involves the removal of one

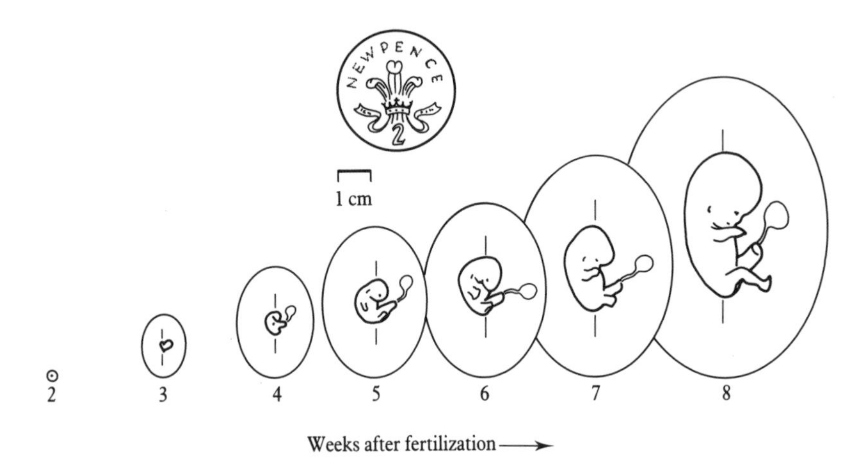

Figure 4 Once the embryo has formed, it grows and develops rapidly. From about eight weeks after fertilization, it is referred to as the foetus. (*Source*: McLaren A., 1984.)

or two cells at the eight-cell stage, or a slightly larger cell sample at the blastocyst stage, and to use that cell sample, the 'biopsy' specimen, to carry out a diagnosis for the particular gene defect for which the couple was at risk (Figure 5). If the diagnosis could be carried out rapidly, fertilized eggs shown to be unaffected could be replaced in the uterus in the same cycle, but if the diagnosis took longer than just a few hours, it would be necessary to cryopreserve (that is, to store in a frozen state) all the fertilized eggs until the woman's next cycle, and then replace those shown to be unaffected.

Of the two options, cell sampling at the eight-cell or the blastocyst stage, the latter offers the opportunity to take a larger cell sample on which to carry out the diagnosis since there are more cells in the blastocyst. A further advantage is that the outer cells, from which the sample is taken, would have started to specialize to form part of the placenta, so none of the cells giving rise to the future embryo would be removed. The biopsy procedure has been applied successfully in an animal model, the marmoset monkey, and does not appear to prejudice later development (Summers *et al.*, 1988), but we do not yet know whether this would be true for the human blastocyst. Another advantage of using the blastocyst stage would be that it need not involve *in vitro* fertilization. The woman could have normal intercourse, fertilization could be *in vivo*, not *in vitro*, and by the blastocyst

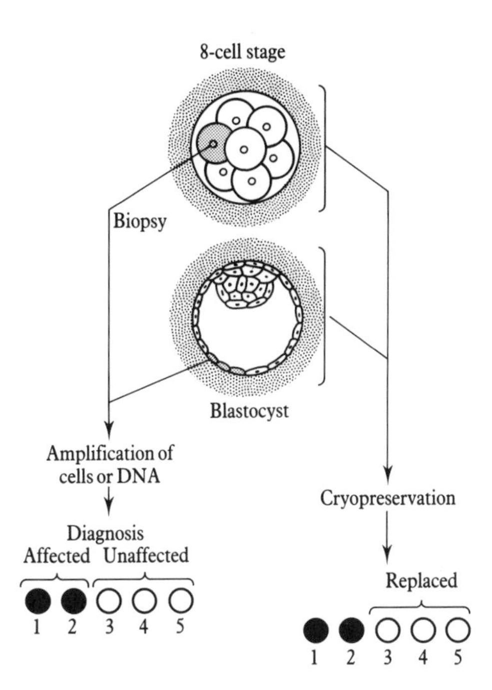

Figure 5 Pre-implantation diagnosis by biopsy could be carried out at either the 8-cell or blastocyst stage (*Source*: McLaren A., 1987.)

stage, when the eggs would be in her uterus, they could be flushed out through the cervix. This uterine flushing procedure, also called 'uterine lavage', has been performed for a different purpose, namely egg donation, in both California and Italy. Unfortunately, flushing seems to yield very few good blastocysts, but those that are successfully recovered give a good pregnancy rate after transfer. Uterine lavage is not an easy procedure and raises both technical and ethical problems, so research on pre-implantation diagnosis in the UK is concentrating on the eight-cell stage.

As an alternative to using cells from the blastocyst stage, the eight-cell biopsy procedure has also been worked out in animal models. Thus, we know from animal experiments that all eight cells at the eight-cell stage are equivalent to one another. If one or two are removed or destroyed, the numbers are made up by extra cell divisions, so that normal young are born.

From the point of view of the woman, either procedure would be relatively non-traumatic. After an initial hormone injection to stimulate egg maturation, an outpatient visit to the clinic to recover the eggs is all that would be required. Uterine lavage can even be performed without an anaesthetic. If all went well in the diagnostic tests, one or two of the fertilized eggs, now known to be unaffected, would be replaced in the uterus, again by a simple outpatient procedure. If the woman failed to become pregnant at the first try, one or two more of the unaffected eggs could be replaced a month later. Once a pregnancy had been achieved, the couple would know that their baby was not going to suffer from thalassaemia, cystic fibrosis or whatever genetic disease otherwise presented a high risk in their case. They would, indeed, be able to feel committed to the pregnancy in the knowledge that it would not be affected.

Diagnostic procedures

In terms of successful diagnosis, the problem with both the eight-cell and the blastocyst procedure is that they yield such minute amounts of tissue – perhaps only a single cell, upon which the whole diagnostic test depends. In order to solve this problem one approach would be to increase the amount of tissue to a level at which standard diagnostic procedures could be used. This might be achieved by inducing those few cells or even the single cell removed in the biopsy to multiply *in vitro*. Unfortunately, it is very difficult to induce cell proliferation at this early stage and the process takes a long time, days or probably weeks, so the eggs would have to be stored. The alternative approach, which in practice is being taken, is to use very much more sensitive methods of diagnosis. I shall now describe some of these quite remarkable new techniques, which can be used to distinguish normal from altered genes carrying a genetic defect. These techniques involve either direct analysis of the DNA of the gene itself or, in some cases, analysis of a product of the gene that faithfully reflects the nature of the gene.

Some genes are expressed (that is, they produce their characteristic effects) at a very early stage of development, either because that is the stage at which their products are used or, in the case of the so-called 'housekeeping genes', because their products are required in all cells throughout life. In the case of genes expressed at this early stage, we could look for the gene product: either the genetic message, that is, a copy of the gene's messenger RNA, or the final protein product, for example, an enzyme. If the gene is defective, the gene product (RNA or protein) will be either abnormal or totally missing.

A sensitive method of looking at messenger RNA is by cellular *in situ* hybridization. This has enabled the message for the human chorionic gonadotrophin gene to be detected microscopically at the eight-cell stage (Bonduelle *et al.*, 1988). At the protein level, ultra-sensitive enzyme micro-assays can be used. A deficiency in production of the enzyme HPRT (hypoxanthine phosphoribosyltransferase) due to a faulty gene causes Lesch–Nyhan disease, a rare but extremely serious genetic disease that kills children in their early teens. It affects boys only, because the gene is on the X chromosome. Using as a model a strain of mice that have a defective HPRT gene, Monk *et al.* (1987, 1988) took cell samples at both the eight-cell and the blastocyst stage and measured HPRT levels by a very sensitive enzyme assay. After samples had been taken, the fertilized eggs were replaced in the uterus of a mouse foster-mother and allowed to develop into foetuses to enable the diagnosis to be checked. In all cases in which there was a low level of the enzyme at biopsy, the foetus turned out to be an affected male; where the level of enzyme at biopsy was higher, the foetuses were unaffected. Thus, it is clear that, even at the eight-cell stage, this enzyme assay identifies the genotype, normal or defective, of an individual fertilized egg.

In mice it is not too surprising that the type or amount of enzyme measured in these studies reflected the genotype of the embryonic cells themselves, since it has previously been shown that the new embryonic genes, as distinct from the mother's, are already starting to be expressed at the two-cell stage (Bolton *et al.*, 1984). Analysis of the pattern of proteins being synthesized at different times after fertilization showed that, in mice, the pattern stays more or less constant up to about 12 hours after fertilization. Between 12 and 15 hours after fertilization the pattern changes as the embryo's own genes start to be expressed and to take control of development. Before that point only the mother's gene products would be seen. The situation is less advantageous in human eggs, where a new pattern of newly made proteins starts only at the eight-cell stage, suggesting that it is only then that the new embryonic genes begin to function (Braude *et al.*, 1988). That means that if the HPRT diagnosis were done for Lesch–Nyhan disease in humans, the results could be misleading, because even at the eight-cell stage it would be primarily the mother's gene products that were being measured rather than those of the embryo.

If pre-implantation diagnosis could routinely be carried out at the

DNA level, not only would it apply to all genes, whether expressed or not, but also we could be sure that we were looking at the embryonic genes, rather than those of the mother.

Polymerase chain reaction – a dramatic breakthrough

Until the past two or three years, identifying a gene defect in DNA from a single cell was an impossible dream and attempts to induce cell proliferation in order to amplify the amount of material for analysis did not look hopeful. Indeed, in 1987 Penketh and McLaren were pessimistic about the prospects of introducing pre-implantation diagnosis to clinical practice within the next few years. But then came the polymerase chain reaction (Saiki *et al.*, 1988), in which the two strands of the DNA of an isolated gene are separated and replicated over and over again in the test tube (see Chapter 3, Box 2). This new technique, which has transformed the situation, depends upon the sequence of DNA to be identified being sufficiently well known to permit oligonucleotide primers to be made that specifically recognize and bind to the boundaries of the relevant portion of the gene. Such information is now available for several genes involved in genetic disorders. With tailor-made oligonucleotide primers and the polymerase chain reaction (PCR), a single DNA sequence, in principle derived from a single cell, can be multiplied a million-fold within a few hours, providing massive amounts of material for analysis. This discovery has removed a major obstacle: suddenly pre-implantation diagnosis has become feasible.

Several genes in specimens taken from mouse eight-cell stages have already been examined in our MRC Unit by enzyme microassay, including HPRT as well as adenosine phosphoribosyltransferase, adenosine deaminase and purine nucleoside phosphorylase (Benson and Monk, 1988); PCR has been used to investigate β-thalassaemia (Holding and Monk, 1989). Male can be distinguished from female fertilized eggs either by enzyme analysis or by PCR. α-Thalassaemia and the collagen-α-1 gene are also being looked at by PCR. All these genes and gene defects are present in mouse models of human diseases.

In humans work is less advanced, but PCR studies are being carried out, with a view to eventual pre-implantation diagnosis, on several genetic diseases, including β-thalassaemia, Duchenne muscular dystrophy, cystic fibrosis (Coutelle *et al.*, 1989), haemophilia and sickle-cell disease (Holding and Monk, 1990).

□ Clinical application

What might be the first pre-implantation diagnosis to be carried out in clinical practice for a couple at high risk of having a baby with a severe genetic disease? It is striking that the pressure to introduce pre-implantation diagnosis quickly is consumer-led, that is, it is coming from couples themselves, and from genetic disease interest groups. In London, β-thalassaemia carriers are a particularly well-educated, sophisticated and organized group; indeed, it was this group of patients, some years ago, that pressed successfully for the introduction of chorionic villus sampling, which is now used routinely all over the world. These patients are now seeking to encourage the introduction of pre-implantation diagnosis, but it is still technically very difficult to identify a single-gene defect, such as the defect in the globin gene responsible for β-thalassaemia, in a one- or two-cell biopsy specimen.

Because the Y chromosome carries a particularly distinctive signature incorporating a number of repetitive DNA sequences, it is not difficult to distinguish male biopsy specimens from female, since the investigator is looking not for a single DNA sequence, but rather a whole series of repeated sequences, which are relatively easy to identify by PCR. Many genetic diseases are X-linked, for example Duchenne muscular dystrophy, Lesch–Nyhan syndrome, haemophilia, and some forms of severe combined immunodeficiency syndrome (SCID). Couples in which the woman is a carrier of one of these sex-linked gene defects are also pressing very hard for pre-implantation diagnosis because, until the gene defect itself can be identified, they are faced with the prospect of aborting all male foetuses, knowing that half would have developed into normal boys. That is an awful decision to have to take.

Professor R. Winston and his colleagues at the Hammersmith Hospital have achieved at least two pregnancies derived from eggs fertilized *in vitro* and identified as female by pre-implantation diagnosis at the eight-cell stage (Handyside *et al.*, 1990).†

Relevant IVF research

The development of pre-implantation diagnosis would not have been possible without a considerable body of past and present research on eggs fertilized *in vitro*. Equally, better prospects for pre-implantation diagnosis will depend upon further research into *in vitro* techniques. Some of the relevant lines of research are summarized in Table 2.

The non-invasive biochemical analysis of eggs fertilized *in vitro* is relevant, not to genetic defects, but to infertility. It involves measuring what the fertilized eggs take up from their surroundings and what they secrete into it, with the aim of diagnosing which eggs are most viable and hence most likely to develop further after replacement in the uterus (Hardy *et al.*, 1989). This would be very important to the success rate of IVF. *In situ* hybridization to messenger RNA is relevant to the determination of the levels of messenger RNA for specific genes. Protein synthesis patterns, enzyme activity and cell surface antigens are all relevant to diagnosis at the protein level. Attempts are

Table 2 Research projects on human eggs fertilized *in vitro* relevant to pre-implantation diagnosis.

Non-invasive biochemical analysis
In situ hybridization to messenger RNA
Studies on RNA synthesis
Protein synthesis patterns
Determinations of enzyme activity
Cell-surface antigen expression
Establishment of cell lines
Embryonic biopsy procedure

† *Editors' note* Pre-implantation diagnosis, for example for Duchenne muscular dystrophy, has recently been established in several hospitals in the UK.

being made to induce cell proliferation and the establishment of cell lines in order to amplify the amount of material obtained by biopsy. Research is also needed into the biopsy procedure itself, for example, to check, as far as can be determined *in vitro*, that removal of one or two cells at the eight-cell stage does not prejudice subsequent development (Handyside *et al.*, 1990). Other research projects are concerned with the possibility of removing tissue from the egg at an even earlier stage: before fertilization. One set of chromosomes is always discarded by the egg in the so-called first polar body, and this offers another possible source of material for early diagnosis (Monk and Holding, 1990).

☐ Regulation of *in vitro* research

All research in the UK on eggs fertilized *in vitro* follows the recommendations of the Report on Human Fertilisation and Embryology, which was prepared under the chairmanship of Mary Warnock. The UK Voluntary Licensing Authority, subsequently called the Interim Licensing Authority, was established following publication of the Warnock Report. The Authority, which functioned very successfully, published an annual report listing all research proposals that had been approved, the names of the scientists involved and where the studies were being carried out. The public availability of this information reassures people, and makes them less likely to think that mad scientists are doing dreadful things to embryos behind closed doors. I am delighted that the UK parliament has now voted to set up a statutory authority that will continue to license and oversee IVF research. The Human Fertilisation and Embryology Authority (HFEA) was established in the autumn of 1990, and took over from the Interim Licensing Authority on 1 August 1991.

The guidelines of the Interim Licensing Authority followed closely those recommended in the Warnock Report. All research proposals were required to be scientifically sound and related to clinical problems. A sufficient number of animal studies should have been done to make it essential to look directly at the human situation. Consent from the donors and approval from the local ethical committee had to be obtained. An important guideline was that any fertilized egg on which research had been carried out *in vitro* should not be replaced in the uterus. So, for example, the licensing authority only gave permission for replacement of eight-cell stages that had been subjected to cell sampling after sufficient research had been carried out on donated fertilized eggs to ensure that, as far as possible, the biopsy procedures would not prejudice subsequent development. The guideline prohibiting the maintenance of fertilized eggs for more than 14 days after fertilization was based on the recognition that some cut-off point was necessary. The primitive-streak stage, 15 days after fertilization, is when true embryonic development begins, and is also the last stage at which monozygotic ('identical') twinning can take place. It therefore marks the beginning of individual development, 'individual' in the sense of 'that which cannot be divided'.

☐ Ethical issues

Pre-implantation diagnosis raises some specific ethical and social problems. Certain problems are associated with any form of prenatal diagnosis: for example, what conditions should be diagnosed? With severe, crippling or life-threatening genetic diseases there is little problem, but what about more trivial genetic defects? Where should one draw the line? And what of more complex characters? Not just single-gene defects but polygenic conditions: for example, an increased tendency to develop heart disease or cancer. And if the number of handicapped babies is reduced, will that lead to increased prejudice against those who are born? This is a concern that is often put forward. All these are problems that need to be considered very seriously.

☐ Pre-implantation diagnosis: a means of avoiding the dangers of germ-line gene therapy

On the negative side, pre-implantation diagnosis is always going to be costly if it involves IVF. On the positive side, it renders germ-line gene therapy unnecessary. It is important to distinguish between somatic gene therapy, the aim of which is to change the genes in the body cells of an individual, and germ-line gene therapy, which is directed at changing genes in the germ cells that are going to contribute to future generations. Attempting to substitute a normal gene for a defective one in a fertilized egg could lead to genetic changes in both the body cells and the germ cells of the resulting embryo. Because germ-line gene therapy would mean tampering with the genetic material of future generations, both the UK licensing authority and all guidelines and legislative codes formulated elsewhere have ruled that it is ethically unacceptable.

Even if it were allowed, germ-line gene therapy would be unlikely to succeed in an individual fertilized egg. Furthermore, if the option of pre-implantation diagnosis were available, germ-line therapy would almost certainly be pointless. Every 'at-risk' couple produces at least as many, if not more, unaffected than affected fertilized eggs. Before embarking on an attempt to cure a defective gene, it would be necessary to discover which fertilized eggs were affected, since it would be undesirable to subject a normal fertilized egg to gene therapy. In other words, pre-implantation diagnosis would be required before the initiation of gene therapy. At that point, the woman would no doubt prefer to have unaffected fertilized eggs replaced in her uterus. Only a very dedicated right-to-lifer would recommend molecular manipulation of the remaining, affected eggs to render them fit for subsequent replacement.

Finally, the most important positive effect of pre-implantation diagnosis would be substantially to reduce the need for abortion on genetic grounds. It could be argued that, in a country that carries out 150 000 abortions a year on non-medical grounds, a few more or less make little difference. From the point of view of society, that may be true, though one has to remember that abortions on genetic grounds make up a high proportion of the late abortions that cause so much concern. But from the point of view of the couple, we know that the advantage of first-trimester, early abortion over second-trimester, late abortion is very great. The benefits of pre-implantation diagnosis, obviating the need for abortion altogether, would undoubtedly be still greater.

References

Benson C. and Monk M. (1988). Microassay for adenosine deaminase, the enzyme lacking in some forms of immunodeficiency in mouse preimplantation embryos. *Hum. Reprod.*, **3**, 1004–9

Bolton V.N., Oades P.J. and Johnson M.H. (1984). The relationship between cleavage, DNA replication, and gene expression in the mouse 2-cell embryo. *J. Embryol. Exp. Morph.*, **79**, 139–63

Bonduelle M.L., Dodd R., Liebaers I., Van Steirteghem A., Williamson R. and Akhurst R. (1988). Chorionic gonadotrophin-β mRNA, a trophoblast marker is expressed in human 8-cell embryos derived from tripronucleate zygotes. *Hum. Reprod.*, **3**, 909–14

Braude (1988). P., Bolton V. and Moore S. Human gene expression first occurs between the four and eight-cell stages of preimplantation development. *Nature*, **332**, 459–61

Coutelle C., Williams C., Handyside A., Hardy K., Winston R. and Williamson R. (1989). Genetic analysis of DNA from single oocytes: a model for preimplantation diagnosis of cystic fibrosis. *Br. Med. J.*, **299**, 22–4

Handyside A.H., Kontagiami E.H., Hardy K. and Winston R.M.L. (1990). Pregnancies from biopsied human preimplantation embryos sexed by Y-specific DNA amplification. *Nature*, **344**, 768–70

Handyside A.H., Pattinson J.K., Penketh R.J.A., Delhanty J.D.A., Winston R.M.L. and Tuddenham E.G.D. (1989). Biopsy of human preimplantation embryos and sexing by DNA amplification. *Lancet*, **i**, 347–9

Hardy K., Hooper M.A.K., Handyside A.H., Rutherford A.J., Winston R.M.L. and Leese H.J. (1989). Non-invasive measurement of glucose and pyruvate uptake by individual human oocytes and preimplantation embryos. *Hum. Reprod.*, **4**, 188–91

Holding C. and Monk M. (1989). Diagnosis of beta thalassaemia by polymerase chain reaction amplification of DNA in single blastomeres from mouse preimplantation embryos. *Lancet*, **ii**, 532–5

McLaren A. (1984). Where to draw the line? *Proc. Roy. Inst. G.B.*, **56**, 101–21

McLaren A. (1987). Can we diagnose genetic disease in pre-embryos? *New Scientist*, 10 December

McLaren A. (1989). Early human development. Why do we need research? *Sci. Publ. Affairs*, **4**, 83–94

McLaren A. (1990). Research on the human conceptus and its regulation in Britain today. *J. Roy. Soc. Med.*, **83**, 209–13

Modell B. (1986). In *Human Embryo Research: Yes or No?* Bock G. and O'Connor M., eds., p. 101. London and New York: Tavistock

Monk M., Handyside A., Hardy K. and Whittingham D. (1987).

Preimplantation diagnosis of deficiency in hypoxanthine phosporibosyl transferase in a mouse model for Lesch–Nyhan syndrome. *Lancet*, **i**, 423

Monk M. and Holding C. (1990). Amplification of a β-haemoglobin sequence in individual human oocytes and polar bodies. *Lancet*, **335**, 985–8

Monk M., Muggleton-Harris A.L., Rawlings E. and Whittingham D.G. (1988). Preimplantation diagnosis of HPRT-deficient male and carrier female mouse embryos by trophectoderm biopsy. *Hum. Reprod.*, **3**, 377–81

Penketh R. and McLaren A. (1987). Prospects for prenatal diagnosis during preimplantation human development. *Baillières Clin. Obstet. Gynaecol.*, **1**, 747–64

Saiki R.K., Gelfand D.H., Stofell S. *et al.* (1988). Primer-directed enzymatic amplification of DNA with a thermostable DNA polymerase. *Science*, **239**, 487–91

Summers P.M., Campbell J.M. and Miller M.W. (1988). Normal *in-vivo* development of marmoset monkey embryos after trophectoderm biopsy. *Hum. Reprod.*, **3**, 389–93

6

Molecular Genetics and New Plants for Agriculture

Professor Richard Flavell

Richard Flavell is Professor of Biology at the University of East Anglia, and also Director of the John Innes Institute in Norwich. Previous to this he worked at the Plant Breeding Institute in Cambridge where he founded the Department of Molecular Genetics. He is an internationally recognized figure in plant genetics, and has been a pioneer in the development and application of biotechnology to agriculture. He has acted as a consultant to various institutions on environmental pollution and tropical agriculture. His recent interests include the molecular analysis of crop plants such as wheat and the potential for their improvement using genetic engineering. He has written over 100 scientific papers and is editor of several major genetics and molecular biology journals.

Molecular genetics research has tended to be concentrated on bacteria and the animal kingdom. There are many who feel that probably some of its most important applications will be in the plant world, and its effects on agriculture and plant genetics might be quite as important as its effects on animals and even humans.

Man has used plants and, in particular, selected variants of plants in agriculture for many thousands of years, but it is only since the discovery of the structure of DNA that we have started to understand the basis of genetic variation. Now we are entering a phase of great scientific significance in plant biology, pure and applied, as we begin to explore and modify plants by genetic engineering. Much of the progress in plant genetic engineering has grown from the discovery that a soil bacterium naturally transfers some specific genes from one of its chromosomes to that of the plant cell that it infects. We have been able to capitalize on this natural gene transfer system between organisms in different kingdoms to insert genes of our choice into some plant species.

Developmental patterns of gene activity have been uncovered and it is now possible to redesign genes to have different patterns of activity, some such as those found in natural mutants or in unique patterns of engineered plants. Some of the most significant changes conferred on plants to date, relevant to agriculture, confer resistance to viruses and insect pests. Flower colours and patterns have also been changed. Examples will be described. The scientific achievements are impressive and we now await when and under what conditions societies will welcome the introduction of genetically engineered plants into agriculture and horticulture.

Plant molecular geneticists, and plant breeders too, are now being forced to analyse their practices and products as never before, because members of the public, in many parts of the world, are questioning the desirability of growing and eating the products of modern plant molecular genetics. There is, therefore, a substantial need to communicate the principles and opportunities of modern plant breeding if society is to gain sufficient confidence in the processes and products of plant molecular genetics to take advantage of all the benefits that they will bring.

It is well-appreciated that over the last 25 years plant science has changed considerably. However, I often like to remind myself, and others, that the most significant changes have come about not because of research into plants, but because of the progress in understanding the biology of DNA and the manipulation of DNA in bacteria.

When Professor Pritchard of the Genetics Department at Leicester University examined me for my PhD degree I was studying micro-organisms. After a postdoctoral fellowship in the USA I made a positive move to leave the busy, intense world of microbial genetics to enter the more pedestrian, lonely world of plant biology. I never thought then that, not so many years later, I would be in a plant science research institute surrounded by several hundred scientists, each of whom grows the bacterium *Escherichia coli*, to produce plant genes, as part of his or her daily routine. Clearly, an intellectual and technical revolution has occurred inside the laboratories.

This revolution is also about to occur out of doors, because plants created in the laboratories are being given properties useful in agriculture and useful to the consumers of agricultural products.

Man is dependent on plants for almost all aspects of his existence, and ever since the beginnings of agriculture, he has been exploiting the opportunities that come about from the reassortment of genes in plants that are the basis of better strains. As an understanding of genetics has emerged during this century, plant breeders have attempted to exploit the new knowledge to get better versions of all the principal crops. Often, this has simply been by the expansion of established, existing processes and methods. But it has often also involved making crosses with primitive ancestors of our crop plants or, indeed, with other species. The plant breeder has always been keen to try out novel and bizarre crosses to get new genes into modern crop species and thus solve certain problems. While the plant geneticist and the plant breeder have an intellectual curiosity to know what might come out of such a cross, there has been a much bigger driving force to the experiments: society has needed better crops and better products to eat, more efficiently produced, and available in much greater quantities.

We are all being increasingly reminded of the exploding world population and its consequences for food production. This in turn demands a sustainable agriculture that does not deprive the soil of essential ingredients or require excessive inputs of chemicals and pesticides. I believe new knowledge from plant science is going to be essential in overcoming some of the problems in maintaining the world food supply as we move into the next century. I am also very confident that some of this knowledge is going to come from research into the biology of DNA and the manipulation of genes as DNA molecules.

What sorts of products do the plant breeder, the farmer, industry, the consumer and the environmentalist want?

Clearly, many different sorts, but disease-resistant crops are always near the top of everyone's list. That is why I will devote much of this chapter to the problem of plant protection and what some of the recent discoveries have to offer in providing new opportunities for overcoming plant disease and pest problems.

□ New genes for the control of viral diseases of crops

Figure 1 shows a field of sugar beet. The plants in the left-hand half reveal symptoms of a virus disease: the plants are not tolerant of or resistant to the virus. On the right-hand side, the plants are growing well and will yield much more because they have been sprayed with an insecticide. The spray solves the problem for the farmer and keeps the yield high. This particular spray does not interrupt the ability of the virus to spread within those plants. What it does is to kill the aphid insects that carry the virus from plant to plant as they feed from the sap. So here you have a disease problem of a virus, but the solution adopted, a chemical one, is to spray with an insecticide.

Viruses cause major problems to crops all over the world, but to a much greater extent in some crops than in others. Viruses, and the insecticides sometimes used to control them when they can be afforded, are often a serious problem. It is therefore no surprise that

Figure 1 A field of sugar beet. The left section shows the symptoms of a viral disease. The other section has been sprayed with an insecticide, which kills the aphids that spread the viruses from one plant to another. Creating plants that are genetically resistant to viral diseases is an alternative to spraying with insecticide.

overcoming these twin problems has been a long-standing aim. Recently some remarkable experiments with new genes have been performed to find alternative ways of controlling virus infections. The story started with the characterization of the molecular details of tobacco mosaic virus. The structure of this virus was established through some very important research; the genetic information of the virus, which in this case is encoded in RNA rather than DNA, was found to spiral through the middle of the virus and to be surrounded by a protein coat that protects it.

From a chemical analysis of all the genetic information, and knowledge of what is encoded, it was discovered that the information to make this protein coat lies in the viral nucleic acid. The discovery enabled a single gene to be constructed in such a way that it would make this coat protein in a form that would be active when placed inside a plant. The gene was inserted into tomato and tobacco plants so that every cell in the plant now made a small amount of the viral coat protein in the absence of the virus. When these plants were subsequently infected by a virulent strain of tobacco mosaic virus, the plants did not readily display disease symptoms – that is, they displayed a kind of immunity.

This result was not predicted, although the experiment had been undertaken because it had been known for many years that, if plants first become infected by a mild strain of a virus, they are often protected against a more virulent, related strain of the virus. This had given rise to the thought that it was some genetic activity of the first virus that probably resulted in the cross–protection. With the technical developments in the 1980s of manipulating genes and inserting them from the test tube into plants, it became a practical proposition to put some individual virus genes into plants and see what happened.

The experiments, as I have indicated, involved finding a gene in a virus, reconstructing it so that it worked in a plant and putting it into a plant. How is this done? Plant geneticists exploited a remarkable discovery, made towards the end of the 1970s, to achieve it. The discovery was that a very common soil bacterium called *Agrobacterium tumefaciens* infects plants and causes a cancerous disease by transferring some of its genes into a plant cell. Here then is an example of a bizarre gene transfer between bacterium and plant that has gone on in nature for an extraordinarily long time. It evolved to provide the bacterium with the selective advantage of a nutritious niche in which to multiply. This discovery was made at the time when the plant geneticist was saying 'It would be marvellous to be able to put new genes into a plant chromosome, and yet we're not quite certain how to do it'.

The obvious answer was to exploit this bacterium, and that was what was done to produce the tomato plants already referred to, and a host of others. The steps along the way involved characterizing the genes transferred from the bacterium to the plant cell. The genes that cause the cancerous growth were defined and removed, taking away simultaneously the disease potential of the bacterium. This did not, however, remove the ability of the bacterium to deliver other genes to a plant cell. So, in the place of the cancer-promoting genes, the genes of interest were substituted. To achieve the gene transfer the plant geneticist takes a piece of plant leaf, or other plant tissue, and infects the outer cut edges with the bacteria. Then the plant tissue is cultured and under the right conditions growth is stimulated from the cells around the cut surfaces. Shoots can be regenerated from the infected cells, and with the appropriate tissue culture procedures a whole differentiated plant that now carries the new gene in every cell can be recovered. In this way, utilizing the miraculous gene transfer system of the soil bacterium, the new process of inserting genes taken from other organisms or synthesized in the laboratory was developed. This ability to change the properties of a whole plant is one of the most significant achievements in the history of plant breeding.

I have given you one example of creating plants resistant to a virus. In the very few years since that discovery was made, about 10 more plant species have been similarly modified so that they are resistant to one of 10 different viruses. In each case the same principle seems to hold: that is, if the plant is making a small amount of the viral coat protein in its cells then, for reasons not yet fully understood, the incoming virus cannot replicate, spread and cause an epidemic.

The first plant to receive a gene designed in the test tube and shown to produce a specific response in the plant was created only in 1983. Now, only a few years later, there has been an explosion in this technology to produce a wide array of plants with different valuable properties.

Monsanto Corporation in the United States has already modified the most commercially important potato cultivar to be resistant to three major viruses found in the US and Europe. The plants are now undergoing trials, and the company believes that it will be possible to release them for commercial use in 1995 or 1996. Now, if those plants turn out to be as commercially valuable as appears to be the case at the present time, it will be a tremendous achievement. It will be of great significance in the history of bringing together plant breeding and genetics research to solve problems that have not been overcome by conventional plant breeding to a sufficient and effective level.†

Other ways of manipulating viral genes to provide disease resistance have also emerged. I do not have space to describe these in detail, but I want to emphasize that in these cases, too, the new approaches began from the knowledge that certain virus strains, which had an extra piece of nucleic acid, seemed to be much less virulent than their close relatives lacking this piece of nucleic acid. This knowledge was the starting point for the plant geneticist to say 'Let me build that piece of RNA into every plant cell before an infecting virus gets there, so that I can give that plant cell the added protection

† *Editors' note* In an entirely different but equally important landmark, Calgene Inc (California) will introduce its genetically engineered rot-resistant tomato into the $3.5 billion annual US market in 1993. In this case, the single gene for polygalacturonase, an enzyme, has effectively been silenced by the genetic engineers; this avoids artificial ripening and increases flavour and shelf life.

that we want'. Consequently, the RNA of the virus was turned into a plant gene and inserted into plants by the process referred to. The expectation was fulfilled and the plants were much more resistant to the virulent virus.

It is interesting to note that these two early advances in plant protection came about because the molecular details of the genes of the viruses were already known, before the gene-transfer system was devised. This, I think, establishes the expectation that, as we learn more and more about genes in plant pathogens and, of course, in plants themselves, new opportunities and surprises will emerge to enable many more improvements to be made by these routes of genetic modification.

New genes for the control of insect damage to crops

Insects are a major problem and devastate crops all over the world. I have already referred to the fact that they are, in many cases, the carriers of viral diseases. In some situations such insects have been controlled by spraying spores of a particular bacterial species, *Bacillus thuringiensis*. These spores contain a protein that is toxic to the larvae of insect species when eaten and therefore provide a biological means of controlling insect pests. Various strains of the bacterium contain proteins toxic to different ranges of insects. Following the advances in gene transfer to plants it became logical to say 'Let's stop spraying bacterial spores. Let's take the gene from the bacterium and put it into the plant, and make the plant synthesize the toxic protein, so that when the insect bites the leaf it will ingest the toxic protein and, hopefully, insect proliferation will then be controlled'. Several species of plants of this type have been created and shown to have a substantially increased resistance to damage by insect larvae of the appropriate species.

So here you see a replacement of one kind of solution – that is, spraying bacterial spores on to crops – by another, a genetic one, involving the moving of a gene and the production of a bacterial protein into the plant. This is a more efficient solution and includes a saving in the energy used to spray the bacterial spores.

I want to refer to another example. It was noted in a world collection of cowpea seeds that one or two lines seemed to be resistant to the devastation caused by beetles during storage. Loss of seed during storage is a very significant problem in Africa. Once these lines had been recognized, it was logical to ask 'What is it about the seed in these particular genotypes that is different, and gives the resistance?'. A piece of extremely good biochemistry, coupled with insect feeding trials, established that a protein in the seed inhibited a particular digestive enzyme of the beetles, thereby producing the resistance. Resistance was therefore due to a natural plant gene.

Why should the value of this gene remain limited to cowpeas, when there is the means of transferring it to many other plant species? This thinking led to the isolation of the gene from cowpeas and its insertion into other crops by the *Agrobacterium* gene-transfer method. The resulting plants showed much less damage due to insect larvae and a much higher mortality rate among the larvae that fed on the leaves.

I began by describing the introduction of genes from viruses pathogenic to plants and then from a bacterium pathogenic not to plants but to insect larvae. Now I have described taking a gene recognized to be valuable in one plant species and putting it into another to gain the required benefit. This gene is being put into cotton, maize and potatoes, for example, by different companies around the world, to help solve the problems of insect attack, and also, hopefully, to help solve the environmental problems associated with insecticide spraying. The latter problems are serious in some situations, especially with respect to cotton, which often requires more than ten applications of insecticide to control boll weevils. The problem is getting worse as insects resistant to the insecticide are selected.

It is, of course, very relevant to ask whether manipulation of a gene that produces a protein toxic to insect larvae should be moved into other plants, crops and food chains. I want to make the point, using the product of this plant gene as an example, that in nature, plant protection against pests and diseases is often achieved via naturally occurring molecules that we would call toxic. There is nothing novel about the concept of toxic molecules in plants. Plant breeders have known for decades that they are manipulating genes that influence the levels of toxins. Therefore, it is inappropriate to conclude that because one is manipulating something that is toxic to some living pest or pathogen, it should not be considered as a means to solving a problem. A much more sophisticated analysis is necessary.

It is sometimes said that it would be very nice to be able to conrol a disease or pest problem in the field by procedures that minimize exposure of man and other organisms to a molecule that is toxic to at least some living organism. Well, that is the beauty of some advances in genetic manipulation, because we no longer have to design the gene in a form that it is expressed in every part of the plant throughout its growth cycle. Now we can exploit the signals in the DNA that regulate gene expression developmentally. Thus, where insect damage in the field, but not during seed storage, was the problem, we could ensure that the gene was expressed only in the leaf, where the insect will bite, preventing accumulation of the toxic protein in the seed that man might want to harvest and use for other purposes.

Genetic manipulation is therefore increasingly giving us the sophistication to design modifications to plants in such a way that we can substantially reduce some problems perceived to result from the introduction of a new kind of protein into part of a plant that is going into a specific food chain.

□ New genes for the protection of crops against herbicides

Another form of plant protection receiving considerable attention consists in making crop plants resistant to a herbicide. Modern agricultural practices certainly need herbicides if they are to be perpetuated with a style and efficiency that many would argue are essential in order to produce food on a major scale economically. Problems are associated with utilizing some of the available herbicides, because of toxic residues left on the soil that enter food chains and because of their toxicity to man when being applied. If we are going to use herbicides, we must be reasonably confident on the basis of exhaustive testing that there are no unacceptable environmental problems.

Herbicides are used to clear plots of weeds before emergence of the crop seedlings. Although helpful in planting and the establishment of crop seedlings, pre-emergence herbicides do not eliminate weed problems during early crop growth. Therefore, selective herbicides that kill weeds but not crops and could be applied after seedling emergence would often be more useful.

Finding herbicides that kill weeds but not crops is difficult. Therefore there is an interest in creating genetic resistance in certain crop varieties. The progress in using genes from bacteria and yeasts, as well as plants, to confer herbicide resistance on plants has been dramatic. This is in part because the growth-inhibiting mechanisms of action of some herbicides are fortunately the same in bacteria and yeasts as in plants. This is a happy, if surprising, fact of nature: that quite frequently the constituent parts of a cell are recognizably the same, whether present in a complex organism or a simple 'primitive' bacterium. Consequently, it was possible first to select soil bacteria that were resistant to herbicide toxicity and isolate the single gene responsible for the resistance. This is much easier to do in bacteria than in plants. The gene might, for example, encode an enzyme that catalysed a step in the biosynthesis of specific amino acids. In the resistant mutant the enzyme was no longer inhibited by the herbicide, but still retained its normal and essential biosynthetic activity. When such a bacterial gene was converted into a form that would be active in plants and inserted into plants via the *Agrobacterium* gene-transfer process, plants resistant to the herbicide were obtained. Later, the functionally equivalent plant genes were used to modify plants to produce herbicide resistance.

In another case in which a bacterial antibiotic functions as a useful herbicide, a gene that detoxifies the antibiotic to protect the antibiotic-producing bacteria was transferred to the plant to confer herbicide resistance. Now élite cultivars of tobacco, tomato, potato, corn and many other species have been genetically modified by the insertion of new genes making them resistant to herbicides. They await commercial introduction into agriculture.

New constraints, legislation and public understanding of genetics

In all of the cases I have described above to illustrate how scientific advances are producing plants of agricultural value, the plants have not been grown outdoors for testing according to conventional practice among plant breeders and geneticists. No, the case for testing them out of doors has had to be argued, often at some considerable length, through local and national expert committees, and formal licensed permission given for release from greenhouses. This is because the plants have been genetically modified by laboratory processes that are different from normal sexual pollination. Now we have laws, or are about to, regulating the release of genetically modified plants out of doors. Is the introduction of these constraints appropriate or is it misguided interference?

Man has been releasing and using crops with new gene constructions for thousands of years. Many of the successful plants growing in our fields now have genetic information from other species that got there only by man's intervention as a geneticist. This was not always done knowingly, or with the same kind of precision as with the new molecular methods, but the exploitation of plants with foreign DNA is certainly not new. I do not object to experts or the general public taking a hard look at what we are pouring into the environment – this is essential and desirable. But I do care that what we can or cannot do should in the end be founded on a strong scientific logic. Because some plants that have been used in agriculture for decades already contain chromosomal DNA segments that originated in other species, it seems illogical to regulate new plants with foreign genes through a different sort of safety legislation. In any case, potentially harmful or undesirable genes can be introduced into an agricultural or horticultural cultivar by sexual fertilization with the same or related species, as well as by fusion of single cells of different species, followed by regeneration of a whole fertile plant, or by *Agrobacterium* gene-transfer techniques. Therefore, to legislate on the basis of the mechanism of gene transfer is illogical. I believe that the evaluation of a plant should be based on its properties and not on the way it was constructed. Of course, the way it was constructed may influence the properties, or potential properties, of the plant.

In attempting to evaluate a plant, we cannot know all its genes and properties. We only have the answers to tests and evaluations we perform, not to all those we do not or cannot carry out. However, from the long history of breeding and agricultural use of our staple crop species we have considerable confidence that the inherent risks involved in creating and selecting new gene combinations are acceptable. The single or few genes we add via molecular and cellular techniques are known and their activities can be measured in great detail. This greatly helps the evaluation process. The plants most difficult to evaluate in terms of risk are those carrying large segments of chromosomes from foreign species that have not been evaluated in agriculture for a long time. Many such plants are not covered by the proposed legislation because they are produced by traditional methods of sexual crossing. In proposing, on the basis of scientific logic, that plants should be evaluated on the basis of their properties, not how they were made, I recognize that I am bringing within the evaluation process all the plants that breeders create, not just those that molecular geneticists create. It would be a terrible and ridiculous bureaucratic nightmare if every genotype a plant breeder produced had to be evaluated and licensed for growth out of doors. However, exclusion clauses, based on experience, could easily be created to prevent breeding and research programmes being halted by ridiculous bureaucracy. Furthermore and most importantly, it is necessary to look ahead to the time when all leading cultivars will contain foreign genes isolated by molecular biology. The number and all the possible combinations of cultivars that will be created by intercrossing will demand that any legislation controlling the growth of plants containing novel genes can easily accommodate the routine activities of plant breeders.

It is entirely understandable and commendable that society should need to be confident that agricultural products are not harmful to man or other organisms. The role of the scientist is to make clear what is involved in plant breeding, what genetic manipulation means and where to look for potential risks to evaluate against potential benefits. It is the scientist who must teach the scientific logic on which procedures and laws should be based. If the scientific community fails to do this and companies say 'We don't want to market plants that were genetically engineered – it will ruin our business, because public perception is not what it ought to be about these plants' we shall be denying the agricultural community and many others the value of extraordinary advances and problem-solving opportunities. We must be completely open about the technology, the science and our own best projections of what genetic changes are doing and how they are doing it. We must get this message across clearly to the policy makers, influential politicians and the people who are going to have a very big influence upon the future of the subject in terms of its application in the next few years. This also means transferring knowledge into the schools and the national curriculum.

I am very confident that in the end scientific logic and, indeed, the demand for problem solving will win out. But they might only win out at the cost of a great deal of argument that will be very detrimental to science and agriculture in the meantime. Therefore, we must find a way to debate the issues and to project the new scientific advances in the context in which they should be seen: that of an understanding of genetics, and an understanding of the products that are created. It is in this context that I very much welcome the opportunity in this volume to highlight the dramatic scientific advances that are knocking on the door of agriculture, the farmer, the food industry and the consumer.

Concluding remarks

I have given you some examples that show that we have seen a revolution in plant genetics built upon the discovery of the properties of DNA and ways of manipulating genes in bacteria. The subject really took off, in terms of its application, in 1983 and we have some tremendous new products on the way. Yet there are problems in releasing these, because of the concerns of society that will not go away.

One of the threads in the debate about the desirability of genetically modified crops is that 'we in Europe don't need such organisms'. We produce many surpluses and our standard of living and health is not limited by deficiencies in the properties of plants. This kind of argument against the development and exploitation of scientific advances ignores the fact that in many poorer countries the situation is different. The problems of food shortages are more acute in most developing countries and any plant improvements with major benefit-to-risk ratios will be essential to sustaining life in these countries as their populations grow during the next century. We should remember that genetic manipulation in plant breeding is a science that has always had a global application and always will. The need to produce more plant material is a global need. We, the rich countries, need to ensure that technical developments are made to help global food production. In deciding whether they are acceptable or not, we should not be parochial but should take a global, long-term view. If we do not, our society will not be promoting the exploitation of the richness of the new plant genetics on which an acceptable future for the planet might depend.

References

Cubitt I.R. (1991). The commercial application of biotechnology to plant breeding. *Plant Breeding Abstracts*, **61**, 152–8

Gasser C.S. and Fraley R.J. (1989). Genetically engineering plants for crop improvement. *Science*, **244**, 1293–9

Knauf V.C. (1991). Agricultural progress: engineered crop species, field trial results and commercialisation issues. *Current Opinion in Biotechnology*, **2**, 199–203

Persley G.J. (1990). *Beyond Mendel's Garden: Biotechnology in the Science or World Agriculture.* Oxford: CAB International

Vasil I.K. (1990). The realities and challenges of plant biotechnology. *Bio/Technology*, **8**, 297–300

13th Report of the Royal Commission for Environmental Pollution (1990). The release of genetically modified organisms into the environment. London: HMSO

7

The Problems of Knowledge

Baroness Warnock

Mary Warnock was Mistress of Girton College, Cambridge until 1992, and has spent much of her professional life investigating the ethics of education and of biology. She chaired the Warnock Committee, which studied the use of human embryos for medical research and gave a guarded but optimistic view of the benefits of such practices for medicine.

The whole issue of experimentation on human embryos is clouded with emotion and requires sound judgements to be made, based on scientific and moral principles. Baroness Warnock expressed her views on this subject at a time when the Human Embryo Bill was under debate in Parliament.

I shall address three questions: Should we pursue knowledge, and the perfecting of new techniques in this field? If, as I think, the answer is 'yes', to whom should this knowledge be made available? These are not medical or scientific but social questions, and we will have to make decisions on such matters as insurance and employment. Wide public participation (based itself on increased understanding) is needed for such decisions to be acceptable. We must seek some kind of consensus.

I have chosen a philosophical title for this chapter, because I hope that, as ought to be the case with anything philosophical, it will lead to thought and discussion in the long run. My purpose is to bring out some very general considerations that might lie behind future thinking on genetics and society.

I shall confine myself to human genetics only. Within this field we can distinguish three different questions. First, should we pursue further knowledge of the map of human genes? (The aim of the map is to discover both how genes are interconnected and how they are expressed.) Secondly, if we do pursue such knowledge, to what medical or social uses should such knowledge be put both with regard to individual humans and to whole populations? And thirdly, if it becomes increasingly possible to know the genetic characteristics of individuals, by whom should this be known?

These three questions may be distinguished, rather roughly, and discussed as though they were quite separate, but it ought to be understood that in practice they are not completely distinct. Knowing and using knowledge are not, in reality, two totally different things. Or rather, acquiring knowledge and using it are not two totally different activities. And if anybody acquires knowledge about anybody else's genes, that knowledge has already become potentially public. Nothing that counts as scientific knowledge cannot be shared. And so the question 'Who else ought to be told?' is already implicit in the concept of knowledge itself.

Yet, though the two go together, the distinction between acquiring and using knowledge in this field has begun to seem, to a certain extent, necessary and also, to a certain extent, acceptable. Hardly anyone would dispute that the rate at which information is being accumulated about human gene systems is prodigious, partly because of the current attempt to sequence the entire genome, partly because of the use of individual gene probes. And this being so, it is virtually impossible, in my view, to say that the accumulation of such information should cease, and that knowledge should be brought to a halt exactly where it is now. There would be something impossibly arbitrary in such a decision, even if it could be universally implemented. And the point is, of course, that it could not be universally implemented. Someone, somewhere, would continue the search, rightly pleading, if challenged, that academic freedom is one of the highest human values, and that to refuse to allow any increase in understanding is contrary to one of the most important moral principles that humans can adopt.

Should we pursue knowledge of our genome?

There seems to be a growing consensus that we must pursue further knowledge. It would be both immoral and, in any case, impossible to call off that pursuit. For example, in December 1989, the European Council of Ministers published a common position paper, in which it was proposed to expend considerable sums of money to contribute to human genome analysis; but at the same time it was agreed that 'throughout the execution of the programme, the ethical, social and legal aspects of human genome analysis should be the subject of wide-ranging and in-depth discussions, and possible abuses of the results or later developments of the work should be identified. Principles for their utilization and control should be proposed.' In this position paper, then, the distinction between knowing and using knowledge seems to be assumed. It was agreed that knowledge, through genome analysis, must continue to be sought. But its use must be further considered.

Before going on to the second of my initial three questions, it is worth considering briefly whether the possession of greater knowledge of human genetics, however well it may be justified, is likely to pose any moral problems in itself. We want to know more, and we should encourage science to help us know more, but will our new knowledge be awkward for us to handle, in itself, whether we use it or not?

The very oddity of this question confirms the artificiality of the distinction between having and using knowledge. Nevertheless I want to raise the odd question: will the knowledge in itself be difficult for us to handle? Many people fear that there will be problems; and indeed they hold that there already are problems. The whole of our language is based upon the presumption that we voluntarily and deliberately do things: I kick, I bite, I scratch, I buy, I sell, I walk and run. I am the agent. We seem to be committed, linguistically, to distinguishing what we do from what is done to us. Moreover, besides using the active and the passive voice of the verb, we can deploy, for example, the optative mood of the verb, in which we say 'I wish that I hadn't done so and so', with the strong presumption that I need not have done it if I had chosen otherwise. I blame myself. I express regret for something that I chose, but need not have chosen, to do. And so we come on to specifically moral language, which essentially accords blame and praise, merit and demerit, for what we do but need not have done, what we failed to do and could have done. We take it for granted, that is to say, in all our thinking, and in all the language in which we think, that our actions can be voluntary, and that the will is free.

But now we are being asked to accept the proposition that more and more of our non-freedoms can be revealed. How are we to adapt our way of thinking? Will not all ascription of responsibility, and all morality fly out of the window? The first thing to say about this is that, though people have raised it as a great new problem, it is by no means new. Philosophers have worried about the freedom of the will for centuries. If it was not one apparent obstacle to freedom, it was another that raised its head. The omniscience of an all-knowing God was at one time thought to be a great difficulty for freedom. Nature was another (for example, to Aristotle). The neurophysiology of the brain has seemed to make freedom impossible, and so, at one time, did the workings of the unconscious. And we have known, after all, about chromosomal disorders for a long time, some of which seem to show that we are less than wholly free to choose our behaviour.

The difference between past and present debates about freedom and determinism is that we could do nothing to change the old obstacles that seemed to stand in the way of freedom, at least not very much. We could do nothing to change the will of God, even if He had destined us for ruin. Though people attempted electric-shock treatment or brain surgery to change people by changing their brains, they seemed hardly to know quite what they were doing, and it did not appear to work very well. Tinkering with the unconscious did not change much either. But now, with the introduction of the word 'engineering', as in 'genetic engineering', and with the extraordinary precision of the discoveries with which we are confronted, as well as the successful alterations that have been made by gene manipulation in the case of plants and animals other than humans, there is a general belief not only that all our actions are programmed to be as they are, and all our thoughts and feelings, too, but that the programme can be rewritten. And so knowledge leads, in this case, directly to fear.

I do not believe, in fact, that there is reason to be appalled by the thought of determinism as such, whether we think of it in terms of neurophysiology or genetics or any other proposed total explanation of human conduct. For I believe that humans can best be thought of as extremely complex physical organisms who have a very quick way of finding out what it is that they want, what it is that they intend, what they think, what they wish, and what they are doing, and that this quick way is roughly labelled 'consciousness'. It is our consciousness that enables us not merely to be aware of our circumstances and environment, but to make our decisions in the light of these circumstances. That *in principle* our decisions might be predictable makes no difference to our own way of deciding. We exercise our awareness and our preferences immediately, without the cumbersome accumulation of scientific generalizations to help us do so. Our freedom *is* our imaginative consciousness of where we are and what we want to do. Humans are so bright that they have devised a way of telling other people, with various degrees of accuracy and truthfulness, what is going on in their consciousness, and from their point of view. However much we knew about the interactions of our own genes and the influence upon us of our complex and changing environment, such knowledge would always be primarily general and only derivatively particular. For example, even if I knew that you were programmed to react with intense emotion to certain kinds of music (and, after all, this is the kind of thing that, long before we knew anything about DNA, we used to recognize with astonishment, as being passed down from one generation to another) I would not know *precisely* what your present state of excitement was caused by. I would not know that it was because you had just been listening to a particular Purcell anthem on your car radio, unless you told me so. It would almost certainly have been impossible for me to predict that you would have switched to that station, that the BBC would have put on that particular programme at that exact time, when I was going to meet you (perhaps by chance). Knowledge of particular facts about people, their intentions, their feelings, and their thoughts (for example, their thoughts about what they have been doing in the last five minutes or five years) must come, and always will come, from the old means, from *conversation*. From our own viewpoint, our way of knowing what our genes are making us do, and how they are determining our reactions is the way we have always had. We have had to recognize our own mental life, and to attempt to make sense of it in the language that we have.

I do not deny that an increased knowledge of genetics and of an individual person's particular genes may make us revise some of our notions, particularly some of our notions about punishment and desert and about the differences between treatment and punishment. But, then, Christianity, for example, might have made us revise our notions on that kind of subject a long time ago. And I do not believe that any amount of extra knowledge can prevent our thinking about ourselves, and thus about other people, as active agents, with plans, intentions, rights and longings that we will try to describe to each other as they occur. Even if in some theoretical sense we are programmed, in practice we shall be acknowledged to be as free as we always were. I do not believe that our new knowledge will radically change our own views of ourselves.

□ To what medical or social uses should new knowledge of genetics be put?

I want now to turn to the second and third of my questions. Granted that we will and we must pursue knowledge of genetics, how should such general knowledge be used in medicine or social engineering? And with whom ought it to be shared, when that knowledge pertains to individuals?

If, as I believe, these are difficult moral and also political questions, then the further question must be asked: how should such matters be controlled, so as to avoid possible abuses, harmful to society as a whole as well as to individuals? These questions are not, or not primarily, medical questions. They are not concerned with how to treat patients, or how to provide them with medical education so as to avoid ill health. Such issues will arise, and doctors will have to answer them, both in general and in particular cases. But primarily, the questions that I am raising are to do with what society will or will not find tolerable, when new techniques are developed. And these in a democratic state are questions for everyone. It is for that reason that I hope to get the issues discussed as widely as possible.

Raising the question 'To what use should genetic research be put?' suggests, of course, that there are, or may be, some kinds of uses to which it should not be put, or perhaps that there are some kinds of uses that would be impossibly expensive, even if desirable. The first of these assumptions is the most important.

People on the whole are afraid of the powers to alter or replace human genes that the new knowledge will probably bring. Some years ago the moral philosopher Jonathan Glover wrote a book entitled

What kind of people should there be? And this question is one that many people think ought not to be asked, or not asked with serious intent. For either they think that God created us as He intended us to be, and each human is valuable as he is, in the eyes of God, or, if they don't use theological language, the very same sense may be detected in what they say. For they think, and they are right, that one of the supremely good things of which humans, as far as we know alone among animals, are capable is of being tolerant of people's differences and charitable towards their disabilities, as well as admiring of their abilities. Therefore, to suggest a political regime, or even a kind of medical practice where such differences might be eliminated and a uniformity imposed, is inhuman, contrary to what is best in human nature.

There is in this respect an enormous difference between Christian civilization and that of the ancient Greeks. The Greeks had no doubt that good people were good specimens of humanity. And if you were a poor specimen, you were *prima facie* worse than your more healthy and beautiful neighbour. If Aristotle had had the ability to determine what kind of people there should be, he would not have been in much doubt. The virtues that they should have would include physical excellence – they should not be sick, they should not be weedy. The word 'virtue' simply meant good quality, of whatever kind. Animals had virtues, so did knives, so did tables. Human virtues were more complicated, but could be listed fairly well by an attentive observer, just as the virtues of a good racehorse could.

Christianity, however, and so Western civilization in general, has taught us to think of humans quite differently from the way in which we think of plants or other animals. So, while we are perfectly happy to root out nettles, improve our roses, or allow selective breeding in cattle or household pets, the thought that some such thing might be practised among humans is abhorrent to us. I am convinced that this attitude is good. It is at the centre of morality. In any case, whether good or not, it is the attitude that we have, and there could be no reason whatever to try to change it. It is not only human, but it is also humane.

However, I believe that there is a duty for all of us, and particularly for the medical profession, to maintain the tradition of improving the life chances of those whom it is within our power to help. And this obligation is just as much a part of Western or Christian tradition as the belief that we are all God's creatures and therefore in some sense worthy of equal respect. Indeed, the obligation to help, to improve life for other people, is embedded in that Western tradition. Think only of the parable of the Good Samaritan.

We may consider the kinds of possible interventions in the field of prenatal diagnosis and germ cell manipulation not as ways of deciding what kind of people there should be, but rather as succour, helping those who are born to live a properly human life, subject to no worse life chances than our own.

Even in the most totally non-interventionist system, not all those pre-embryos that come into existence will actually continue to live. The medical profession entering the field of prenatal diagnosis can think of itself therefore as doing no more than ensuring that the survivors have the best possible chances. Similarly, if, in the future, gene therapy becomes a reality, as it surely will, then the medical profession can think of itself as doing no more than is already acceptable, that is, replacing a defective part by one that is healthy and will lead to health. There is nothing absolutely magic about genes: they are parts of our bodies, just as our livers and kidneys are.

Two crucial issues arise at this point. The first is the distinction between somatic cell manipulation and germ-cell manipulation. It seems to me perfectly easy and realistic to regard somatic cell therapy, the replacement of a cell from an embryo by a better cell, as simply part of what doctors have always been accustomed, indeed, bound, to do; namely to work for the good of an individual patient. In the case of germ cell therapy the patient will not yet have been born, but it is an *individual embryo* who is the particular patient of the doctor in this case. The techniques are more sophisticated than of old; but we are accustomed to increasingly sophisticated techniques hardly less amazing, such as the possibility of organ transplant.

If, on the other hand, we think of the manipulation of the DNA of germ-line cells, that is, eggs or sperm, doctors cannot tell who their patients may be. Furthermore they cannot realistically predict the effect, long term, on society of changing the genetic make-up of generations of unborn humans. On grounds of ignorance and unpredictability alone, there should be extreme caution about any such intervention. Indeed, I am of the opinion that germ-line genetic manipulation should be absolutely banned, and by law.

The public at large may well become accustomed to the notion that genetic therapy is a new and sophisticated way of helping a particular human, provided that the decision is taken to develop gene therapy only in relation to single, somatic genes. If the medical profession insisted on manipulating genes in a manner that would affect unnumbered future generations, the public would, rightly, be afraid of the consequences, and would, through their governments, insist on the prohibition of all gene therapy. Rather than press for germ-line gene therapy, we must try to get the principle of prenatal selection of embryos, where families are at risk, thoroughly accepted. If such progress were made, then the need for germ-line intervention would no longer exist in the long term and the question could be allowed to lapse. I believe that the law could and should be invoked to prohibit all germ-line genetic manipulation, and that this is the first and most important regulation that must be introduced.

The second issue is perhaps more easily settled. Is genetic intervention (and I am now talking only of somatic cell intervention) justified in the case of all genetic defects, or only of some? The identification and location of single genes responsible for genetic diseases is progressing very fast. It seems to me that in order both to spread the knowledge of what is possible, and to allocate resources sensibly, there ought to be an agreed list of priorities, setting out where society wishes to put its resources, or at least its major resources. This would cover priorities both in treatment and in research. For the time being, of course, there is not much difficulty. There is no shortage of severe, horrible and lethal single-gene-related diseases, where research, and, in due course, treatment, would be appropriate, and where everyone will agree that medical research and practice should be funded. In the case of such diseases we must learn to see gene therapy as simply a subclass of ordinary therapy. We need not raise questions about the future of society as a whole. For the life of an individual (and thus, of course, in the long run, and indirectly, the life of society at large) would be infinitely better if that individual did not suffer from that disease.

There is, as far as I know, no philosophically wholly satisfactory way of comparing decisions about the good of an individual (or his or her interests) with the good of the world at large (or global interests). The demands of the particular and the general come at us quite differently. For example, one might try saying to a particular infertile couple 'Don't worry about your infertility, accept it with gratitude; there are too many people in the world as it is'. It would be a rare person who would react favourably to such an approach. The two issues do not seem to be comparable. In any case we cannot expect individuals to make or seek to implement decisions of policy. Such decisions are public, not private, and are in general political. There are many cases where private interest seems to conflict with political or social good. But in the case of individual gene therapy, there is simply no need to raise this kind of problem. We have to think how best to alleviate individual human suffering. And this, after all, has always been the preoccupation of the medical profession, even of those of its members who are concerned with public health issues. The long-term future of the world at large need not come into it.

We have to find out the conditions that cause the worst suffering, and that offer the most optimistic outlook for therapy, and put these conditions at the top of our list of priorities. Somewhere along the line will come a calculation of cost, but the main imperative is to publish a list so that both the medical profession and the lay public may know what is possible, what is worth doing and what is permitted. In 1987 the Australian Health and Medical Research Council issued guidelines in this field. The Council stated: 'The choice of disease is critical. Initial trials should be limited to diseases which cause a severe burden of suffering and for which there is no treatment.'

☐ To whom should knowledge of our individual genetic make-up be given?

Before addressing head-on the question that was my third at the beginning, namely, who should have access to information about the genetic analysis of an individual, there is a preliminary question connected with it. Should it be compulsory for any individual to undergo genetic screening? These two questions are, as I say, plainly linked, because compulsion would seem necessary only where the information derived from the analysis was to be passed on, at least to some persons or agencies other than the patients themselves. I do not pretend to be able to give answers to any of the questions that I am about to raise. Indeed, it is on them that I hope discussion in future may focus, because it seems to me that some sort of consensus, and then, very possibly, regulation, will be needed in this field. The subject matter is so new that there are no great lines of tradition to which we may turn in trying to settle the question of moral rights and wrongs here. We have to start more or less from the beginning.

Let us start with the relatively simple case of diseases that are known to be genetic in origin, and where a family cannot be unaware of the presence of the gene or genes in some family members, for example, haemophilia or Huntington's chorea. In such cases, the question of compulsory screening may well arise. It is already the duty of the medical profession to give such family members genetic guidance, and to tell them what their options are if they want to have children. It is essential that doctors should take on this duty, or see that it is properly carried out by somebody else. It may be essential for women to know whether they are carriers, and for men who are symptom-free to know whether they, nevertheless, carry the gene. If, as I sincerely hope, pre-implantation screening, selection of embryos and subsequent *in vitro* fertilization become more secure and available for affected families, then, even before gene therapy is a real option, there are practical steps to be taken, and most people will want to make use of them. The case of Huntington's is perhaps more controversial, since its symptoms do not appear until adulthood. There might be people who did not want to know their own genetic make-up. If such persons lived entirely alone, were self-employed, uninsured and had no intention of having children, then it would be very easy to say that no one should compel them to acquire knowledge that they would prefer not to have. Compulsion could be justified only where other people were concerned. But, of course, other people always are concerned. And this is where the question about compulsion spills over into the question: who should know?

This question becomes far more complex when no one can state precisely, as they can in the case of, say, Duchenne muscular dystrophy, what is the probability of a boy's actually developing a genetic disease, or of a girl's being a carrier. The gene analysis of an individual is certain to reveal that he or she has some genes rendering that person more liable than the average person to suffer from some disease or other. He or she may, for example, have a higher than average likelihood of suffering a coronary attack, or of becoming schizophrenic. Or, of course, of suffering less disastrous conditions like hay fever. Do individuals have a duty to find this out about themselves through genetic screening? And if they have this duty, for whose sake do they have it? Who has to be told?

The more refined and accurate such analyses become, the more it may seem that there are decisions to be made. In the old days, for example – in my day, I mean – on getting married one might say to one's future husband 'Well, our children are bound to be short-sighted. Look at us.' Or 'They're sure to be asthmatic.' But, on the whole, such reflections did not stop one either marrying or having short-sighted or asthmatic children. The question of possible invasion of privacy or of a possible right not to have anyone know about one's genes comes only when the knowledge is both relatively accurate and may put one at a disadvantage, compared with other people, if it becomes public.

There are four main areas in which the question of whether we have a right to privacy must be raised. The first is within the family. The second is within educational establishments. The third is at work, and the fourth is in the area of insurance. (There may be other areas, but these are the most obviously important.)

Within a family (and I would include siblings among the family as well as potential spouse and children), it seems to me that on general moral grounds there ought to be openness. But such openness must, I suppose, be a moral imperative that cannot be enforced by law, but only encouraged by reflection and advice. However, if openness in the other areas were to be enforceable, then openness in the family could scarcely fail to exist as well.

In the case of educational establishments, a great deal turns, I believe, on whether the condition in question is one that is degenerative. In such cases schools or colleges must be told. It also turns on whether certain environmental factors may make the genes more likely to result in illness. This is absolutely crucial, for example, in the case of schizophrenia, which often manifests itself first in people between the ages of 18 and 21, when they may be at university. I do not know whether there are environmental circumstances, for example, undue stress, or loneliness, to be avoided by people known to be liable to schizophrenia. If there are, and if there are environmental arrangements that one can make to avoid or mitigate these circumstances, then it seems to me just as important that universities should know about that liability in any of their students as it is to know that any of them is coeliac or epileptic. There are already too many avoidable tragedies among students that have arisen because the university or college has not been told about possible or incipient mental illness. It is time people stopped being afraid that no institution would accept a student if a liability were revealed. We ought to make up our minds as to whether mental illness is indeed an illness, or whether it is something else.

The fear that someone would be disadvantaged in getting jobs if their gene analysis were known is even more potent, and probably more justified. To decide how much information a prospective employer is entitled to demand is a social question, raising issues of equality and justice that are certainly not capable of being decided by doctors or scientists, but must be addressed by employers and employees, that is, by society at large. My instinct, but it really is no more than this, is that, once again, we need an agreed list of medical conditions, such as might prove actually dangerous, about which employers might be entitled to ask, rather as, in applying for a driving licence, everyone has to state whether or not they suffer from epilepsy. Questions about probabilities and proneness to certain diseases should not be permitted unless there are some very important environmental precautions that ought to be taken, and that a particular job would rule out. If there were such a limitation on what it was legitimate for a prospective employer to ask, this would have to be enforced by law. An analogy here would be the privacy to which people are entitled with regard to former criminal offences or prison sentences.

The issue of insurance is perhaps the most intractable of all. On the whole, the principle guiding an application for insurance is that of utmost honesty, and it seems essential, if the business of insuring is to continue, that that should not change. If insurers began to demand a genetic analysis, they could penalize, through their premium, anyone who refused to have such an analysis or to reveal its contents, or they could simply refuse to insure such a person. At present it looks to me as though the insurance companies might have the whip hand, making impossible any kind of right to privacy or to ignorance that one might feel inclined to protect. Certainly in the United States there is evidence that, if you try to conceal something from an insurance company, it will come out in the end. It is already extremely difficult for some groups of people, thought to be at risk, to get insurance of any kind in the United States. This, again, is not an issue for doctors, but for society at large. It is extremely important that we come to a fair and equitable decision, even if it means that all of us pay higher premiums for insurance. For we must avoid a situation in which there are some people who find it impossible to get insurance of any kind, these being the very people who are likely to need it most. This is an issue that it would be well for society to decide in advance, by a consensus between people who thoroughly understand the insurance business: actuaries, doctors and lawyers.

There is one final question that is not one of my three, and which I shall simply raise, without attempting an answer. It is sometimes asked to what extent people have a right to enrich themselves by patenting the outcome of genetic research. I am inclined to think that, though this is a question bearing very much on the whole issue of the funding of medical research, it is not unique to the knowledge of genetics, but relates to scientific knowledge as a whole. It is a matter for public decision, but on a broad front.

By way of conclusion I will try to sum up what I have said. I have argued that society must permit, and indeed, encourage, the pursuit of knowledge in the field of genetics. I further argued that the use of that knowledge in medicine should be subject to regulation, or at least to guidelines where the principle should be that of attempting to remedy disease to produce an individual healthy person, rather than to produce something quite new, a superperson. The use of genetic manipulation, that is, ought to be regarded as an extension of the use of other techniques in medicine, and should be guided by exactly the same principle, that of benefit to the individual. And I argued finally that the most difficult questions before us are those about the sharing of knowledge. On these questions it is urgently necessary that a consensus should be sought and that legislation should be finally introduced, resting as far as possible on that consensus. Only in this way can society's natural fears of our new knowledge be shown to be baseless.

Index

K

L

M

N

O

P

R

S

T

U

V